W9-DCV-603

9/77

MORRIS FRASER, born in Scotland and educated in Ireland, is Senior Registrar in Psychiatry at the Royal Belfast Hospital for Sick
Children.

Children in Conflict

Children in Conflict

Morris Fraser

WITH A FOREWORD BY ANTHONY STORR

Basic Books, Inc., Publishers

NEW YORK

Copyright © 1973 by Morris Fraser
Foreword copyright © 1977 by Anthony Storr
Library of Congress Catalog Card Number: 73-92721
ISBN: 0-465-01043-1
Printed in the United States of America
77 78 79 80 81   10 9 8 7 6 5 4 3 2 1

Contents

77892

Foreword

Great Britain has a millstone round her neck. Its name is Northern Ireland. Since 1969, when violence exploded there, murder and destruction of all kinds have become endemic; at the time of this writing, no end can be seen to the conflict, in spite of all attempts to restore law and order and repeated political initiatives to bring the two sides together.

What has been the effect upon the children of Ulster of being exposed to and involved in so much violence? *Children in Conflict* is a remarkable, shrewd, and compelling study by a psychiatrist who has treated many of the disturbed children from the areas of worst conflict. It is also extremely well written.

Moreover, the significance of this book extends far beyond the unhappy country of Ireland. For, although the roots of the Irish troubles lie far back in history and may be thought unique, the kind of conflict that goes on in the streets of Belfast is common in many parts of the world, and the tensions that turn Protestant against Catholic and vice versa tend to occur between any two groups of men with different cultural traditions who are living in close proximity.

Politicians, understandably enough, are suspicious of psychiatric interpretations of political situations; but I must urge them to read this book, since the conflicts that they are attempting to resolve take their origin from deeply rooted prejudices that, generation after generation, have been implanted in the minds of children. It is only when these prejudices become diminished that there will be any hope of a final end to conflict in Ulster.

It is easy to forget how long the violence has gone on. Children now at school have never known streets free of armoured cars, bullets, petrol bombs, stone throwing, broken glass, and the perpetual threat of death. Children of eight and upwards participate in violence—hurling missiles, making bombs, setting traps of wire for armoured cars. Most appalling of all, because the soldiers are reluctant to fire on them, children are often used as front-line combatants, hurling bombs in situations into which adults dare not venture, acting as a protective screen behind which their fathers, armed with rifles, can take both aim and shelter.

Dr. Fraser delineates two varieties of disturbed children, which are so common as to be typical. Ringleaders of violence are generally boys in their early teens whose fathers are either interned or active members of the IRA. Such boys are violently hostile, not only to the security forces, but also to their own parents and to all authority figures. At the opposite extreme is the child of ten or so, commonly a girl, who suffers from night terrors, incontinence, and all the usual symptoms of intense anxiety.

As psychiatrists would expect, but others may not realise, the children who display severe psychiatric symptoms are those who are exposed to disturbances in their homes as well as in their country. Children are so adaptable that, even in conditions of war, they do not become mentally ill unless the adults in their immediate environment are themselves mentally disturbed. The most frightened children are those whose parents do not control their own anxiety; the most aggressive are those whose parents act out their own aggression. Again, as psychiatrists would expect, anxiety symptoms in adults, as evidenced by the number of prescriptions for tranquillisers, are worst in areas where conflict is dreaded but has not yet actually broken out.

Dr. Fraser is surely right when he affirms that "the Ulster conflict is not a holy war." He gives convincing evidence that the Catholic ghettoes of Derry and Belfast have formed in exactly the same ways as the black ghettoes of Harlem and Watts. In other words, the conflict is not between equals professing different ideologies, but is the revolt of a minority who have become scapegoats for an insecure, and therefore intolerant, majority.

The pejorative phrases that, in America, white Anglo-Saxon Protestants apply to blacks are word-for-word the same as those that Ulster Protestants apply to Catholics. They "breed like rats"; their estates are "rabbit warrens"; they are "subhuman," "lazy," and so on. "Daddy says we're poor because the government takes his money to give to the

Catholics and all their children and priests to keep them in luxury because they don't work. . . ." This, from an Ulster boy of eleven.

The Catholics are the untouchables of Ulster, and, like their counterparts in India and Japan, are scapegoats—hated, despised, and yet, at the same time, feared. Dr. Fraser reviews the economic and other factors, now familiar to students of politics, that lead to the revolt of the ghetto. These are the same in Ulster as elsewhere: rising prosperity in the surrounding areas combined with frustration in the ghetto when a hoped-for sharing in this prosperity is disappointed. Revolt does not take place in the absence of hope.

Dr. Fraser, through his studies of individual children and his studies of ghettoes elsewhere, has amply established that in Ulster, as in the southern states of America, emotional prejudice is so deeply rooted that political solutions alone will not abolish it. His principal prescription for Northern Ireland is that the present total segregation of Catholic and Protestant in schools be abolished. In other parts of the world, it has been shown that integration of different groups at primary school age does in fact lead to diminution of prejudice. It is more difficult to think of Catholics as rats if, as a Protestant child, you have habitually learned and played with them.

In some ways this is a depressing book, for it underlines the desperate depth of the Ulster conflict. There are no short-term solutions, yet the hope that prejudice can ultimately be abolished is not wholly vain. Dr. Fraser's book is a notable contribution toward the eventual reestablishment of peace.

Anthony Storr F.R.C.P., F.R.C.PSYCH.

Consultant Psychotherapist, Oxford Area
Clinical Lecturer in Psychiatry, University of Oxford

Introduction

Misery has a thousand disguises. But it is never to be read more clearly than in the face of a child, black or white, Irish or Vietnamese. Stark terror, too, is as unmistakably present with a Negro youngster fleeing before a segregationist mob as with a dirty-faced Belfast urchin, his back to a barricade, facing steel Saracens with his last weapons—half a brick, or a rag stuffed into a milk-bottle dripping with cheap petrol. Children's tears speak plainly in any language. Or do they? The main task of this book is to explore the ways in which adults and, especially, children in very different cultures behave under a variety of acute stresses.

It is only four years since a physician wrote to me from Vietnam: 'The war has been going on for so long that no one under 35 has any real memories of what things were like before. The children think that "peace" is a beautiful time and place where all problems are solved—something akin to heaven on earth.'

Here, at least, is cold logic. With clinical detachment, the doctor condenses the dual human tragedy of Vietnam to two brief, pithy sentences. The second tragedy, though no less than the first, at least follows understandably from it. But it would have seemed impossible then that a study of children under stress would shortly have to focus predominantly not on South-East Asia but on Ulster—a tiny, tortured community of a mere one and a half million stateless men, women and children. Why Ulster?

War is variously an art, a field for heroes, a game for dogs, a 'biological necessity'. But to most ordinary families it is nothing more nor less than a mischief foisted on them by remote politicians and generals. It has its terrors, but also, for civilians at least, its compensations—the heightened sense of community, the legendary war-time spirit with its washing fluttering bravely along the Siegfried Line. But stress takes on new and horrifying meanings in a war where there are no civilians—where children and their parents are themselves the combatants, drawn into conflict by powerful forces with vested interests in a deeply divided community.

How powerful? Look at it like this: it is easy enough, when nations are at war, to train fighting men to unload bombs on a darkened city from fifty thousand feet. But it isn't so easy to produce a man who can walk into a crowded restaurant with a bulging briefcase and look into the faces of mothers and children whose limbs he will shortly scatter across six streets. It isn't so easy to teach young children not only to bombard an Army scout car with rocks, but also to jeer and stone the ambulance that carries away the body of the young driver. Yet it can be done: this last incident has just taken place at the time of writing. In Ulster, children have been used as never before in guerrilla warfare, and their training has to start early. Later in the book, I have written about what these children learn in the strange school that is an Ulster ghetto.

But this book is not, of course, one simply about Ulster children; the implications of the past five years' events in the province extend far beyond its almost ridiculously small area. For one thing, as I have pointed out in Chapter 3, the Irish Republican Army, far from being a new phenomenon, was in 1916 the first in a long line of modern guerrilla movements extending from the Irish rising of that year, through the Russian and Chinese Revolutions, to the present Vietcong, Tupamaros and PFLP campaigns. Now, with bigger prizes in sight, there is every sign that the cycle is restarting, again in Ireland, largely in urban settings, and on a much more formidable scale. What happens in Ulster today will probably happen in England tomorrow and in the United States the day after—given the same conditions. In this book I have interpreted conditions largely in terms of childhood experience, which is paramount and universal.

The first chapter holds the substance of a kind of waking nightmare that has haunted me (and others) for the past four years. It is the result of countless interviews with children whom I have encountered during this period, both as a child psychiatrist and as a leader of a youth organisation in one of Belfast's last 'mixed' areas. They spoke to me individually and in groups, at home, at school and in the street. All conversation within quotes is strictly verbatim, but I have interspersed some newspaper quotations as an occasional link with reality as others saw it. The need to employ this technique will, I hope, be clear.

The next two chapters, taken together, centre on the growth of racial and sectarian ghettos in cities. They are set mainly, though not entirely, in Ulster, with emphasis on the influences, both mundane and bizarre, that have surrounded children growing up there. It is far from easy for an outside observer of the province to disentangle the social pressures and deprivations following the guerrilla warfare from those that caused it. In this section I have taken a stance, broadly, somewhat before the outbreak of violence in 1969—perhaps when the present terrorist leaders were growing up. (It is not, I think, widely enough appreciated that the IRA is as much an Ulstermen's army as its Protestant counterparts.) The third chapter ends with a brief account of the events that led directly to the rioting in Belfast and Londonderry in 1969, the opening skirmishes of Ireland's latest and bloodiest revolution.

The following chapters, 4 to 6, tackle the thorny problem of the link between wars, riots, and mental illness. Chapter 4, dealing with adults, concludes that this link is rather anomalous, and looks for reasons. Chapters 5 and 6 are about what happens to children under the worst kinds of stresses and after the most gruesome kinds of experiences that can be imagined. What, for one thing, are the consequences of an entire childhood spent behind barricades? Again Ulster is in the forefront of my examples. (Here, children who were born since the start of the 'troubles', and have thus never known peace, are just starting school.) I have reviewed material from a wide variety of sources, in the hope that the conclusions I would reach could be of use to teachers, parents and others who are faced with the psychiatric casualties, young and old, of disaster.

The remaining chapters contain the kernel of the hypothesis on which much of the book is based—and may also explain why the earlier subject-matter ranges, with apparently undue freedom, from Belfast to Los Angeles, from street battles in Derry to race riots in the southern States. The conflict in Ulster is essentially a racial one, in part inbuilt, in part artificially created. An Irish Catholic who joins the Government or the security forces is, to his compatriots, an 'Uncle Tom'. To Ultra-Protestant newspapers, Civil Rights marchers are 'niggers' and 'apaches' (Chapter 7). This section explores the way in which the doctrine of the white Negro is practised by both adults and children. Then, drawing on evidence from racial ghettos outside Ireland, mainly in the United States, I have written about the social and economic pressures that make for a collective retreat to racial or sectarian enclaves and, within this context, the myths and mysteries of youthful aggression. The last two chapters, a search for solutions, again take Ulster as the starting-point, but I have tried to reach conclusions that will be useful wherever there are to be found children in conflict.

Finally, I have to make it clear that, while I gratefully acknowledge information and advice from numerous sources, all views expressed are mine alone and do not necessarily represent those of colleagues. All names have been altered, as have other details that might identify adults and children described in the text. This, of course, excludes those adults whose names are, by their own choice and at least in Ulster, public property. It is by no means irrelevant to this book's subject that the Irish still go lusting after strange gods.

I

Fact or Fantasy?

The boy in the green anorak took my pencil and carefully drew two parallel lines on the back of an envelope. 'That's the street, right?'

He added a neat row of dots outside each line, then a rectangle in the middle. 'These are the lamp-posts, and that's the Army Land Rover coming up the street. You tie your cheesewire between two of the lamp-posts about six feet up. There's always a soldier standing on the back of the jeep; even with the searchlights, he can't see the wire in the dark. It's just at the right height to catch his throat. Then, when the jeep stops, we can come out and throw stones.'

His friend, two years older, disagreed. 'Only kids throw stones. What *we* do is to fill our pockets with them and carry hurley-sticks. If you put a stone down on the ground and swing the stick as hard as you can, you can hit a soldier below his shield and cripple him. We once cut a squad of thirty-six down to six in ten minutes like that.'

The third boy supported him. 'Then,' he said, 'we come in with the petrol-bombs. If a soldier lowers his shield to protect himself against the stones, you can lob a bomb over the top and get him that way.'

I pointed out that wire-breakers were now welded to the front of all Army vehicles. One passed as I spoke, but the three boys on the pavement didn't trouble to glance at it.

'We know all about that,' said one contemptuously. 'Look.' He drew again. 'One each side, you carry the ends of the wire along the pavement to two *more* lamp-posts. If you didn't do that the first two posts would simply be pulled out. And we use a double thickness of wire. The wire-breaker can't really snap it—so if you don't get the fellow on the back you can at

I

least stop the jeep. It's a sitting target for the others then.'

'How many others?' I asked apprehensively.

He shrugged. 'As many as we can get. Not the kids, of course. We give them the bin-lids. Their job is to stand at all the corners and bang the lids when they see British soldiers coming.'

'Not your job?' I ventured. He gave me a withering look. 'I'm nearly twelve,' he said icily.

It was time for a change of subject. 'What else do they teach you at this club? It *is* a club, isn't it?'

'It's a kind of youth organisation. It was formed just two months ago. It's called the Young Boys of Ireland and there are about a hundred in our branch. We learn how to fight soldiers and prods [Protestants]. And how to make petrol-bombs and all.'

He showed me where they met—in a hall above a billiards club—and said, 'They give us all whistles like police-whistles. When there's a riot we blow them and lead the soldiers and police down the side-streets, where the lads with the guns are.'

'A kind of ambush?'

'Yes.'

'Is this a junior branch of the IRA?' I asked.

The boys exchanged a quick glance. One said yes, another said no. There was an awkward silence, then the third boy volunteered, 'Mr Brown, who gives us our Thursday night lectures, is in it, though.'

'We have uniform too,' said the first boy quickly. 'Black berets for official occasions, like funerals, and green berets for—well—field work.'

I pointed to a triangle of green protruding from his anorak pocket. 'Is that the green one?'

'Good Lord, no—that's my Scout beret.' He looked at the church clock and whistled. 'We're late! Come on!'

Jamming their berets on their heads, the Falcons Patrol stampeded across the road.

In the long, brightly-lit hall the smoke and darkness of the New Lodge Road seemed as remote as Cambodia. The Scout Leader gave the order. The flag ran to the top of the pole; the Falcons saluted with the rest, then ranks broke for British Bulldog.

Brian was fifteen and growing fast, protruding from his uniform at all points. At the end of the meeting he handed his beret to the Scouter with a wry smile. 'You'd better have that. You can pass it on to one of the kids.'

'Too small for you?'

'Oh no, it's not that. It's just that I can't wear it.' The Leader's mounting puzzlement cleared as he explained. He pointed to the left. 'I can't wear it down there because the Army will think I'm one of the Provos[1] and pick me up. And'—he pointed right—'I can't wear it up *there* because of the Junior Orangemen.[2] They know that we've got Catholics in the Troop, and they call me a Fenian-lover. I said that in the Scouts we didn't make any difference, and so now I daren't go out in the uniform; they'd be waiting for me. I don't mind for myself so much, but they would follow me home and put the windows in. As it is, I'll have to take the long way round.'

'Not your lucky night, on the whole.'

'Oh, I wouldn't say that.' Brian fumbled in his uniform pocket and produced two tenpence pieces. 'I found these on the way here, in a pool of blood. I lifted them out and cleaned them. There must have been a fight —or maybe they belonged to that policeman who was shot.'

He held the coins out for inspection, then noticed the expression on the Leader's face.

'Oh, don't worry,' he said. 'I'm only giving you the beret. But four bob's four bob.' He went through the swing doors and into the street.

Outside, Brian was approached by two of the younger Scouts. They asked him to walk home with them. Brian asked why.

'These two fellows came up,' explained one, 'and asked if we were Protestants. I said yes, so he took us into an entry and took a Union Jack out of his pocket and told us to kiss it or else.'

'Did you?'

'Think we're daft? Of course. But there's more of them down the street. We're frightened.'

'Come on, then,' said Brian. They left the friendly circle of the one remaining street light and went off down the New Lodge Road, close to the wall.[3]

Divis Street, the main route to the Falls area, must be the most depressing thoroughfare in Europe. There are no windows—as the term is generally understood—only bricked-up or boarded-up rectangles. At rare points a gaping hole has been overlooked by the Sappers, and filthy curtains are still visible between jagged edges. The only vegetation grows in thick tufts from senile chimney-pots, and the only water is running out from below a door, evil-smelling effluvium from a damaged sewer-pipe. It has been flowing for so long that the pavement around is green and slippery. The wind has piled litter into bizarre snow-drifts, and, at a deserted bookie's, the burglar alarm has been ringing unheeded all morning. Nearby a shop is on fire; tall flames are flowing upwards into a low pall of smoke. There are no police or firemen; nobody pays any attention.

Further up, at the main intersection, the traffic lights are broken again, felled like trees to support last night's barricades.[4] In the middle, standing on the remains of a burnt-out lorry, four small boys all around the age of ten are directing the traffic with stern efficiency. A heavy Army troop-carrier comes to a halt at a signal from the youngsters, then roars off along the Falls Road at another.

Further up still, the Belfast Child Guidance Clinic, although not very far away, is at least comfortable. Certainly, Paul seems to think so. When he comes to see me he stretches out his legs to their fullest extent—admittedly not very far, since he is only thirteen—and rests his head against the armchair's tall back. His referral to a psychiatrist has followed a long series of escapades, grouped appropriately on his last charge under the head of 'riotous behaviour'.

'Got hit with a rubber bullet,' he said proudly on his last visit. 'I've got a bruise, a big black one.'

'What happened?'

'A soldier got me down by Divis Flats. It's gone all blue today, and yellow at the edges.' He showed me a bandaged wrist. 'Got clubbed too.'

He said: 'The first barricade in our street was last August; it was the best one in the Falls, built with bricks so that no one could get through. We spent a whole week building it. There was just one small peephole. But if the Army hadn't come we would have all been burnt out by the Protestants just the same.'

'Yet now you go out and throw petrol-bombs at the soldiers?'

It was a question, and he answered it patiently.

'Look—we needed them then, but not now. Now we're armed and ready for the Protestants. We have to get rid of the Army first. I go out and fight them almost every night, and I get the other kids organised. The smallest ones that can't fight can do things like carrying messages—that's how they start.'

A much larger organisation had grown from a corps of messenger boys at Mafeking. But Paul would not have appreciated the parallel, so I let it go.

'Do you really think that you, with your stones and petrol-bombs, can get rid of the British Army? What's the point in it, anyway?'

'But—these are *British* soldiers—here in Ireland . . . !' He looked at me helplessly.

'You have to do your bit?'

'That's right,' he said with relief. 'And we'll get them out in the end. They starved them out of Dunkirk, didn't *they*? *We* can starve them out.

'How? See if anyone gives them a cup of tea or anything? It's a petrol-bomb round them the next day. Or a tarring and feathering.

'A petrol-bomb's easy to make. You fill a milk-bottle with petrol and use a rag soaked in paraffin as a fuse; petrol is better. You have to watch out not to throw it too soon. If you count three first, then it bursts when it hits him. They nearly got me once with a baton-charge, but I got over a wall just in time.

'The Army are even worse than the police or the Protestants; they think they're tough. They rough up people for just looking at them. And they're unfair. They only search Catholic houses, never the others. They arrested my uncle once; he died of a heart attack later, lifting concrete blocks to make a barricade.

'Junior IRA? No—I'm a full member, in the Official branch, not the Provisionals. The Provisionals want to use violence; we use peaceful means.

'Petrol-bombs? Well, you have to defend yourself somehow. They've got big nets now, you know; they're going to fire them from a big bazooka. The net comes down on you, then they run over the top and trample you.

'It's just nonsense that only two soldiers were killed last night. We saw six lying dead in the Grosvenor Road. They had been lying on their stomachs shooting until we threw a nail-bomb over and they were blown up. We knew they were dead because I turned one over and his eyes were kind of staring up at the sky.

'But you have to watch out for the Provisionals because they believe in violence.'

Bernard is eleven and small for his age. His hair is cut short, his face is newly-scrubbed, and he is dressed in tidy clothes and a flowery tie. He swings his legs and chats freely.

When I asked him whether his home district was quiet he said:

'It will be peaceful until four o'clock.'

'What will happen then?'

'That's when our school gets out. Then we show the Army something. Stones, petrol-bombs, nail-bombs and all. We learn how to make them and all.'

'Who teaches you?'

'Big fellows,' he said vaguely. 'And they can get jelly [gelignite]. You know these perspex shields are no good against petrol-bombs? But you have to throw a few first. Then the burning petrol melts the perspex and the soldier has to drop it and run. If you're quick then, you can get him with another. And if you put in soap, like Fairy Liquid, it helps it to stick.

'Paint-bombs are good too. You can throw one at the windscreen of a jeep when they're going at speed, then they can't see.

'If you're lucky they crash,' he added wistfully.

I asked him about current Press claims that boys were being paid by older men to throw stones, and he laughed.

'If they were paying, the whole street would be out. But they'd be wasting their money, because we do it for nothing.'

William wears a chunky sweater with red, white and blue stripes. He told me he was ten. I estimated nine and later found from the records that he was eight and a half. He told me that when he grew up he was going to join the Army, and hoped for a posting back to Belfast.

'The Army are needed here to keep the Catholics down. It was Bernadette's fault starting all this; she's a Fenian and should be burned to bits. All the Catholics should be killed or burned. They shoot peelers [police-men] and Protestants, but I'm in the Junior Orange Lodge and we know what to do with them.'

His friend, who was thirteen and looked eleven, was more specific.

'We have steel spars that we get from Metalwork in school; we sharpen them. We are keeping the ends sharp for Fenians. We are working on an idea so they can be fired out of a bow, and we practise on dummies.'

I asked whether he knew any Catholics.

'Some used to live at the end of our street. We never bothered one because she was only an old woman. They've left now. They go whining to the Cor-poration and pretend they've been threatened.'

Brendan, a stocky youngster, preferred to draw rather than talk. He had selected three chalks from the box—a white, a yellow and an orange. He drew three parallel bands of colour with bold strokes, enclosed them in white, and added a flagpole. Underneath he printed the letters 'The Flag of Ireland'.

'That's why we hate the soldiers,' he said proudly. 'They're British, but we Catholics are Irish, that's why. We wave an Irish flag at them. I've thrown stones at them often.'

'Who started all this?'

'The Protestants. They threw stones at the Civil Rights and started it. It was Paisley shouting at them. We need the Army to protect us from the Protestants or they would burn us out. But we hate them. We burned a Union Jack once; that was great. We stretched it be-tween two goal-posts and set it on fire. The soldier near done his nut. I think I'll draw another Irish flag now. A bigger one.'

Fact or fantasy? The transition, when it came, occupied only a few months . . .

Belfast Newsletter, 30 October 1970

At eleven . . . Joe is an expert in making and throwing petrol-bombs —and he is a deadly shot with a half-brick . . . He has injured fully-armed British soldiers and is thought to have lamed one policeman. He has led his gang against the organised ranks of troops in vicious street warfare . . . He and his gang have had the strange experience of forcing the soldiers . . . to retreat before their barrage of stones and petrol-filled bombs.

Belfast Telegraph, 5 February 1971

Television audiences have been shocked by boys of four and five lobbing bricks and stones at the troops. Some of the children, unable to throw in a more sophisticated and forceful manner, toss the missiles 'underhand' as they would were they throwing a stick to a dog.

Daily Sketch, 8 February 1971

Rioters again ordered their children into the front-line battle for Belfast yesterday . . . Five children, all aged under twelve, were detained last night, bringing the total of youngsters held to more than 60 in the past three days.

Daily Mirror, 8 February 1971

. . . a boy of 14 had his hand blown off by a bomb he was about to throw . . .

Daily Telegraph, 8 February 1971

Groups of IRA men are now known to have paid sums between 15s and £3 to small children to go out and pelt soldiers with missiles while they shelter behind doorways with automatic weapons.

Daily Express, 9 February 1971

He was no bigger than my own six-year-old. He . . . teetered to a halt three yards from the soldiers. The street was a perilous carpet of stones, broken bottles and jagged metal. The soldiers . . . watched as he swung his arm in the glare of a burning single-decker bus . . . The half-brick fell harmlessly at the feet of a battle-weary corporal . . . The kid retreated, picking up another rock . . . He trotted back to his pals. Nine-year-olds, 12-year-olds, 14-year-olds. The little ones are assembling the piles of ammunition. The big ones are hurling it.

Belfast Telegraph, 24 February 1971

At the sight of a Police Land Rover . . . 'the brats', as the RUC refer to the youngsters who run around the streets like ants, reach in a reflex

action for stones and bricks. Ritually, they bombard the vehicle as it passes, then chase after it, still hurling stones . . . 'Every time we take it out it depreciates by 50 quid,' said the driver.
Sunday Mirror, 27 February 1971

. . . a gang of thugs paced up and down the kerb where Catholics live apart from Protestants . . . There were twelve of them in all. And the oldest must have been about 13 years old. Most of the others were about eight . . . the mini-terrorists . . . the child with a bottle in his hand and an obscene curse on his lips . . . bottles by the score . . . fragments of paving-stones which they could hardly lift . . . They pelted us with their own fire-bombs, the blazing fire-lighters they looted a few minutes earlier from the shop . . . Nobody tried to stop them. Nobody dared.
Belfast Newsletter, 11 August 1971

Two boys, aged about 11 and 14, were seen carrying a bomb to the front of the Army Recruiting Centre . . . The boys ran off and seconds later the bomb exploded . . .
The Standard, Windsor, Ontario, September 1971

His face betrays no emotion as he speaks but his eyes are dark and dis-tant, as if troubled by the memory.

'I was out at our door and the Protestants were coming at us,' he says. 'They were coming down Bombay Street and shooting machine-guns. You could see the bullet holes they made, and they were burning people up. I saw a man, his stomach was hanging out. They put a hanky to his stomach to try and stop the blood, and I just felt I could kill the Protestant who done it.'

In a country that has been torn apart by more than two years of bitter bloody sectarian warfare, such gory reminiscences have become common-place. But what sets this recollection apart, and at the same time typifies the utter tragedy of the fighting in Northern Ireland, is that the warrior who witnessed this scene—and has since been subjected to others just as bloody —is a boy of 13.

While most children . . . around the world . . . are chattering about things like football and rock music, worrying about things like exams and dates, the children of Ulster are being reared as tough, street-savvy young fighters—a generation of street urchins living in a perpetual, chaotic state of imminent violence, schooled not in the three Rs but in the morbid vocabulary of 'gelignite' and 'nail-bomb'.
Belfast Telegraph, 1 October 1971

A boy aged about twelve planted a bomb which wrecked a . . . shop in Belfast last night. Detectives have interviewed an eye-witness who

saw the boy climb from a car with a deadly smoking parcel.
Daily Express, 11 October 1971

Troops under nail-bomb attack in Belfast yesterday held their fire. For the two nail-bombers were boys aged about eight. One of the bombs exploded ... The other, which failed to go off, was retrieved by one of the boys who ran into a house.
Daily Mail, 28 October 1971

The IRA is sending children armed with Tommy guns into the streets of Belfast. Yesterday two 13-year-old boys pumped bullets at an Army patrol ...
Belfast Newsletter, 18 August 1972

Children aged between ten and 17 caused the death of a soldier in Armagh last night in one of the most horrific incidents during the present Ulster troubles. The soldier was killed by a crowd of children who stoned his armoured car, hit him on the head with a brick and caused the vehicle to crash. An Army officer said he was sickened by the actions of the children who, when the ambulance arrived, continued the stoning and injured two policemen.

The language may be emotive, but it is understandably so. Massive deployment of young children in street guerrilla battles became, from 1970 onwards, a prominent feature of the disturbances throughout Northern Ireland; it is a new and terrifying twist to an already tragic situation, one virtually without precedent in modern history.

Perhaps the word 'riots' can be disposed of at this stage. Even after the four-year siege of Derry's Bogside and the gradual, but equally conclusive, battering into the ground of most of West Belfast, the term still suggests outbreaks of violence which are largely unplanned and sporadic. The shorter expression is used here for convenience, but my own feeling is that its terms of reference are too narrow. 'Urban guerrilla warfare' more accurately describes a situation of prolonged and bitter territorial dispute with continuous armed confrontation at boundaries. And, with increasingly sophisticated guerrilla techniques, there is now a role for even the youngest members of the warring communities. At first, children carried messages and warned of approaching troops—usually by banging bin-lids against a wall or pavement. Later, whistles were issued and used, and a further strategy was the use of a flexible wooden plank which, when stood on at one end, lifted at the other and then allowed to spring back against the pavement, produced a convincing imitation of rifle-fire. This was found to be a useful decoy. Soon, as rioting spread from Derry to Belfast, children were making petrol-bombs and passing them forward to their elders in the front line. I have often seen them do this, then retreat hurriedly into doorways and alleys.

But within the first year the situation changed. Now the children, rather than the adults, are usually in the forefront of onslaughts on the security forces. They throw stones and petrol-bombs at fully-armed soldiers with an inexplicable disregard for their own safety—they use firearms and gelignite. What has happened? And what kind of children are these?

Outrage, disbelief, anger jostle for space in the Press leader columns.

Belfast Newsletter, 6 February 1971
In these children . . . all the characteristics of the anarchist are there. Soldiers are . . . skittles at a Belfast street corner . . . targets for every hate-inspired youth who cares to inflict injury on them.
Daily Express, 9 February 1971
. . . tragic pawns . . .
Daily Mirror, 8 February 1971
˙ . . . most atrocious horror of all . . . young children . . . in the forefront of the attack on the troops . . . Unimaginable! . . . a bloody shambles.
Daily Sketch, 9 February 1971
Just kids? Forget it . . . 12-year-olds can be as dangerous as adults . . . Everywhere you go through the bigoted, blood-stained streets of Belfast, there are the kids, screaming and pelting the British soldiers . . . hooched up on hate . . . prancing like urchin dervishes through the murderous cross-fire . . . How many more kids have to be killed and maimed before the parents, the priests, the parsons and the street-corner politicians of Belfast come to their senses?
Sunday Mirror, 28 February 1971
New-style terrorists . . . Family discipline has gone in the same direction as truth, compassion and honesty—to Hell . . . If Belfast has a problem on its hands in 1971, God knows what it's going to be like in 1981 when these kids have grown up.
Newsweek, 19 April 1971
The brawling children of Ulster . . . have passed prematurely from the innocent games of childhood to the deadly serious business of street warfare.
Daily Mirror, 9 August 1972
How can anyone repair the damage done to the hearts and minds of the children of Ulster? The parents who see them going berserk must surely discern that this is the Irish tragedy. It is profoundly tragic that the children of Ulster can no longer be called the innocents.

Belfast Telegraph, 16 August 1972
   Mere children . . . expendable instruments of terror.

Although scathing, these commentaries would of course be perfectly fair
if they were based on a balanced view of the entire problem. But they are
not. The Ulster child problem is much more than one of hooliganism, of
vandalism and naked aggression. Behaviour which makes 'good' television
may be only a small part of a child's total response to a terrifying situation.
In fact, I believe that it is. It is easy, on television and in Press photographs,
to show *behaviour;* that is the nature of the media. But it is almost impossible, on the same media, to show the fears and stresses that provoke the
behaviour. It is, at best, a lopsided portrayal—the response without the
stimulus. Children, too, are natural exhibitionists; they are more than willing to pose, military-style, in helmets and battle-dress, with toy guns, sticks
and shields. No one, it seems, has ever thought it worthwhile to point out
that pictures like these could be taken where children play in any city in the
world. But coming as they do from Ulster, they are commonly used, naïvely,
as vehicles for impassioned comment.[5]
   The point I want to propose here is that this 'revolutionary' activity is
only *one* end-result of a long series of frightening social and environmental
pressures. There are other responses.
   Consider this, from a Belfast mother. Her child, aged ten, lives in a block
of high-rise flats in the Falls area.
   'She hasn't been outside the flat for weeks now. The last time, we went to
her granny; then someone started shooting across the playground at the soldiers. Since then she hasn't wanted to go to her granny's again.
   'She went down to the door last week, but she saw a foot-patrol coming;
she screamed and had to be taken back inside.
   'I could hardly get her to come for her appointment today. All the way
here in the taxi she sat with her hands over her face. She wouldn't take
them away. She was shaking all the time. I don't know what to do; I am at
my wits' end.'
   Patricia herself added: 'It's worse now—now that the new soldiers have
come. They have these big visors, and you can't see their faces. It's . . . oh,
it's awful. They call them the Death Brigade. If you put your head out of
the window they're always looking up and when they see your head they
point their guns straight at your face. We're on the ninth floor but we can't
sleep at night because they rev their engines up all the time.
   'I don't like Daddy going out at all because I think he will be shot or
interned. They took my friend's father last week and he had done nothing.

Last night they came and took some men away. When they were leaving they shouted that they would come back tonight for the rest.'

Janet, aged twelve, said: 'I can't sleep at night for thinking about fires and burning. There are five of us in the bed, and I sleep nearest the window so as my small brother won't get hit with a bullet. Every time I hear a loud noise I shake all over.'

Peter, also twelve, told me: 'Last month [August 1971] we all went to the refugee camp in County Wicklow. It was great because we were away from Belfast; we all cried when we had to come back.'

These extracts from clinical interviews suggest, rightly, that children in Belfast react in two distinct ways to the experience of street violence. The aggressive response is, of course, the most publicised. However, of equal concern to a psychiatrist is the mounting number of children who suffer severe and lasting emotional damage as the result of being exposed, at first hand, to riot conditions. In many cases these nervous disturbances have persisted long beyond the time when the child's own area has become quiet. I am still seeing children suffering from fainting fits, asthma, epilepsy and hallucinations, symptoms clearly precipitated by the stressful events of August 1969 and after. Many of these young patients are handicapped to the extent of being unable to go outside their homes to play or go to school.

But these children, in contrast to the 'mini-mobs', have been forgotten by almost everyone except their parents. And there are other myths to be toppled, other ghosts to be laid . . .

A Canadian wrote to me: 'It is absolutely tragic that these children's parents actually encourage this behaviour. It's sickening.'

An American: 'I have been saddened and upset more than I can say by what I have seen of these young hooligans and vandals on TV. I just can't put my feelings into words.'

An Englishman wrote: 'They are simply mindless young thugs. The sooner Northern Ireland floats away and sinks in the middle of the Atlantic, so much the better for all decent people.'

Why did they write to me in particular? Well, perhaps it is natural that the people most deeply concerned will turn, if not always rewardingly, to a child psychiatrist working in the area for some answers. As it happens, very many journalists, parents and welfare officers do just that. There are basically only a few questions. First—what has been the effect of continuing violence and tension on the mental health of children and adults? Will there be long-term effects? What drives young children to violent, almost nightly confrontations with a fully-equipped army at very real risk of death and in-jury, and why is nothing being done to prevent them? *Can* anything be done? And finally, there remains, unasked, the question that English and American enquirers are much too polite to put in so many words: 'In God's name, what kind of a society can produce children like these?'

It was a sense of this deep perplexity that stimulated the research and observations on which this book is based. The following chapters will attempt, in some measure, to answer these questions, in particular the third one, because the children I have quoted are not freaks or isolates. Their stories are disturbing not because they are unusual, but because they are typical.

2

The Matrix

Matrix is a useful word; it suggests just the right degree of environmental contribution to individual growth and development. Certainly, where quality of environment is in question, it is a less ugly word than 'ghetto'. But this term is now impossible to avoid or ignore; in the communications media it has become as firmly wedded to Belfast 1970 as it ever was to Warsaw 1940 or Lodz 1942. We have Protestant ghettos, Catholic ghettos, sectarian ghettos and, even less happily, ghetto areas and a ghetto mentality. It has become part of the stereotype. Berlin has its wall, Glasgow its slums, Belfast and New York have their ghettos. But is this usage justified? In the search for an answer we first have to scan, fairly broadly, the images, influences and experiences that surround an Ulster child as he grows up. Later in the book, some influences which seem to me of major importance will be examined in more detail.

The word 'ghetto' is Italian in origin and is believed to derive from the name of an iron foundry in Venice beside which a large Jewish community grew up in the early fifteenth century. The third and fourth Lateran Councils, largely to restrict Jewish economic activity, had prohibited Jews from residing amongst Christians, and had enjoined that Jews wear a distinctive badge. Later, following a Papal Bull of 1555 which enforced Judeo-Christian segregation, all Italian Jews were removed to ghettos in Venice and Rome. Other Jewish ghettos of this period were in Fez (Morocco), Frankfurt and Prague. Looting and intimidation in these Jewish sectors were common, and for a Jew to venture outside the ghetto was to invite victimisation or bodily harm.

The European ghettos were opened, for the most part, in the wake of the

French Revolution, but the practice of segregation was revived by the Nazi government before and during the Second World War. In Warsaw, Jews were free to enter the Aryan area and to attend to their normal business during the day, but at night the ghettos were closed by barbed wire and armed sentries. These communities, in contrast to those of the earlier period, had no internal autonomy and were effectively governed by the SS guards.[1]

The definition of a ghetto has broadened somewhat since 1940 to include situations where immigrant groups and Negroes (especially in the USA) have been forced into segregated areas by common social and economic pressures.[2] But the word, for example as applied to Belfast, is still highly emotive; as well as suggestions of overcrowding, poverty and squalor, it carries an immediate implication of political junketings, corruption in high places, and the planned segregation and harassment of a minority.

In so far as it suggests conscious motivation on the part of government, the use of the ghetto model may not take sufficient account of how Belfast came into being—of its sprawling, pseudo-organic development virtually independent of political influences. The city's early population patterns, in fact, derive mostly from its geographical position.

As the opening of a long, fertile corridor, it was shelter to only a few small farm settlements before the Norman conquest. But in the twelfth century, as a ford across the muddy estuary between the conquered lands of Antrim and Down, it was the scene of many a bitter struggle between the Anglo-Normans and the native Irish. Gaining at last an uneasy foothold at this strategic point, the Normans built a castle and, probably, the first wooden boats.

A bridge was built across the Lagan in the sixteenth century and, after the last fight with the Ulster chieftains in 1595, the town was captured by Sir John Chichester, the first lord deputy of Belfast under Elizabeth I. After this the 'plantation' of North Down and South Antrim by English and Scottish Presbyterian settlers went on apace.

Civil commerce, especially with the Continent, flourished after Cromwell's victories, as, a little later, did the shipbuilding and linen industries. Belfast was a natural outlet for the export of raw and manufactured materials from the hinterland as well as from the embryonic city; English and Scottish merchants built large houses on the northern side of the river and, on the south side, where shipbuilding flourished, rows of Protestants' dwellings grew up in the Ballymacarret area. The Shankill Road came into being as the settlers built upwards from the level sloblands past Peter's hill (where Unity Flats now stand), and the Sandy Row district was the area

where the first mill and foundry workers, also Protestants, lived. Further
south, the lush, wooded farmland of the Malone area was cultivated by
wealthy English farmers, and still retains its upper middle-class aura.[3]

Thus, until the mid-eighteenth century, Belfast was almost exclusively
Protestant. An estimate of 1708 put the number of Catholics in the city as
'not above seven'. It was not until Catholics, attracted by the new industries, began to move in significant numbers from the south and west
that their own areas began to be defined—the Falls Road, growing
south-west from the old Fall Mill, the Andersonstown and Ballymurphy districts, damp farmland by the Bog Meadows that had been rejected by the
English colonists, and the Ardoyne and Short Strand districts, on the peri-
phery of the prosperous Shankill Road and of shipbuilding Ballymacarret
respectively. The pattern of segregation in the city was already being laid
down. And by 1782, under the twin disadvantages of the Penal Laws and of
having the poorest land and jobs, the Catholic areas soon degenerated into
the worst slums. The first of a long series of riots was soon to follow[4] and, by
1922, housing had been firmly reinforced along sectarian lines.

So it might be said that the divisions of Belfast were virtually built in, a
function of geology as much as theology—something born of the soil itself.
In a broadcast talk of 1967, the then Professor of Geography at Belfast University offered a subtle analogy. He said (referring to the strata on which
the modern city stands): 'I like to think of this colourful geological succession as symbolising the successive culture layers that have contributed so
much to the personality of the city. The overriding basaltic flows, looking
like burnt porridge turned to stone, lie at the southern end of an igneous
province that extends far to the north through western Scotland, and they
stand for the dominant Scottish element in the population. The chalk, geo-
logically an isolated outpost of the English lowlands, symbolises the English
settlement of the seventeenth century, the relatively thin but rich filling of
the sandwich. Below are the native clays and marls, a pliable but unstable
foundation, liable at times to cause landslides, representing the Irish element.'[5]

Even now Belfast, as a city, has no corporate identity; it is still a 'settlement' town, a cluster of village communities. There is not even a simple
descriptive word for a native. You can be a Londoner, a Glaswegian, a
Liverpudlian or even a Dubliner, but who has ever heard of (say) a Belfas-
tian? The great Shankill–Falls divide is, of course, well-known, and with
reason. In each area, even before the street disorders and massive population migration from 1968 onwards, the dominant religious group, Protestant or Catholic, outnumbered the other by 95% to 5%. More than a year

before the districts were separated by a heavily-guarded tangle of barbed wire and steel, a survey demonstrated minimal interaction between the two populations. They patronised different shops, supported different football teams, read different newspapers, even walked on different sides of some streets and waited at different bus-stops.[6] But apart from all this—apart from the fact that a Shankill housewife would more readily contemplate shopping in Dublin than on the Falls, apart from the nightly confrontations at the Peace Line, convenient for this purpose in an Orwellian fashion —Belfast is still intensely parochial. Each community, or tribal area, centres round its own church, school and shops. The priest, the minister, the headmaster hold immense influence, each in his own area; an expedition to the shops of central Belfast is, for the ordinary resident, something of an un-usual event. (For the time being, we are leaving aside the upper and middle classes, disparate groups in any community, and concentrating on the work-ing class, about three-quarters of the population, the group invariably thought of when ghettos are mentioned, the group that has to bear the brunt of civil unrest and the conditions that precede and follow it.[7])

So what of living conditions in general? Did Belfast, before the outbreak of disorder in 1969, resemble any other working-class city—Richard Hoggart's working-class culture[8] or Donald Horne's 'northern metaphor'?[9] In many ways, yes, but the points of difference are crucial. To set a socio-logical description of Belfast alongside Hoggart's monumental study of urban life is a most illuminating exercise. The healthy, productive commun-ity is compared with the rapidly ailing one and, by addition and subtraction, some factors emerge with harsh clarity—dissensions, failures of adaptation between groups, economic pressures, fears, attitudes and aggressions that finally led to the end of normal life in Belfast and Londonderry in 1969. While I am primarily dealing with Belfast, most of the results I am going to examine also apply to other Ulster cities; important differences will be indi-cated where they occur.

Arriving at Belfast from the airport, your first view of the city is from the elbow of one of the nastiest bends imaginable, where the main route from the north turns downwards towards the Shankill Road. From here you can see how modern Belfast, like many another industrial city, lies in an elong-ated bowl at the estuary of a river, the river dredged, moulded and channel-led alongside the urban area to long, straight fingers extending deep into dockland. The city has often been said, particularly aptly, to resemble a stranded lobster. There are high hills all round, to the point that traffic exits, especially to the north, are very limited in number, making at times for intense traffic congestion. From your high viewpoint it will also be evident, even on the clearest day, that the Clean Air Act has yet to make inroads from the suburbs: the central area, deeply recessed, is at times almost hidden by layer upon layer of thick smoke—blackest where it belches from the oil-refinery and the ships in the estuary, bluish and luminescent around the cranes and gantries of the docks and the great steel skeletons of the shipyards. Harland & Wolff's yard has a total work-force of some nine thousand; the firm is Belfast's biggest single employer. The firm is Protestant-dominated; the five hundred Catholics employed are definitely in the minority, but relations, at least in modern times, have been fairly harmonious.

Immediately around the shipyards, in East Belfast, is the wedge-shaped Ballymacarret area. It is almost exclusively Protestant, and membership of the Orange Order is virtually total in the male population. The adjacent Short Strand and Cromac areas are equally strongly Catholic. Between the two districts, deeply set among trees, is the heavy bulk of St Matthew's Catholic Church, around which sinister tales of snipers in the tower and arms in the vaults have for the last two years revolved and become current among the Protestants of the area.[10] Although these reports have been disproved by Army searches, fear and anger remain, and the church is still heavily flanked with twisted barbed wire and sentry posts.

But, as has already been hinted, religious polarisation reaches its maximum and best-known expression in the existence of 'the Shankill' and 'the Falls', and the glorification of each area with the definite article—a distinction probably last accorded to the Crimea. English newspaper sub-editors have not infrequently dismissed the phrases as colloquialisms, struck out the definite article, and added 'Road'. But each is in reality an area consisting not only of the central artery, but of a huge, interlocking maze of much smaller streets, of microscopic terraced houses, shacks, yards and alleyways, dun-brown, row upon spidery row. The Falls and Shankill Roads run parallel in a westerly direction; the Crumlin Road and the Ardoyne area, further north, are a terrible no-man's-land, the scene of almost nightly rioting and gun-battles. Now, all the streets linking the Falls and Shankill areas are sealed off by the Army—many by steel stakes driven permanently into the ground, some even by high brick walls. This barrier, which winds uninterrupted for three-quarters of a mile, used to be known officially as the Peace Line, and unofficially as the Freeland Fence (after the then General Officer Commanding) but the joke has worn thin, and neither phrase is heard now. Still, the ugly reality remains, and may well become permanent.[11] Press reports concentrate on the civil unrest, on the rioting, as the cycle waxes and

wanes—but all the year round now there are appalling rush-hour jams as the traffic squeezes, inch by inch, through the streets still left open, over the ramps, past the armoured cars at every street corner and, at city school playgrounds, alongside the tangle of barbed wire below and the soldier in his eyrie above. Every schoolboy can now identify and give you the exact specifications of a Saracen, a Saladin, a Whippet . . . Any youngster too, hearing an explosion, could give you a fair estimate of the charge of gelignite used—'That was a fifteen-pounder.'

But these are manifestations visible only since 1969, merely the outward signs of divisions and hatreds that have been growing up in West Belfast for many years previously. Until 1969 this area, at least superficially, differed little from Hoggart's typical northern English city.

'. . . smoking and huddled working-class houses . . . Recognisable styles of housing—backs-to-backs here or tunnel-backs there . . . Houses usually rented, not owned.

' . . . depressing, massed proletarian areas; street after regular street shoddy, uniform houses . . . mean, squalid and in a permanent half-f study in shades of dirty-grey, without greenness or blueness of sky . brickwork and the woodwork are cheap . . . the terraces are gap with sour and brick-bespattered bits of waste ground . . . lowering between the giant factories . . . goods lines . . . gas-works . . . ch pubs. Rough, sooty grass pushes through the cobbles; dock and n on a defiant life in the rough and trampled earth-heaps at the co waste-pieces, undeterred by 'dog-muck', cigarette packets, old elder, dirty privet, and rosebay willow-herb take hold in some or in the walled-off space behind the Corporation Baths. A night the noises and smells of the district—factory hooters, the stink of the gas-works—remind you that life is a mat clockings-in-and-out. The children look improperly fed clothed, and as though they could do with more sunlight

True, but Hoggart points out that while to a visitor depressing, to the resident they are home, his natu insider, these are small worlds, each as homogeneous village.' Much more unsavoury, dirty and odorifero of multi-storey flats—notably Unity Flats and Divi in the past twenty years. The harsh concrete cliff multi-coloured facings, nor do a few drooping sl areas look any less like prison exercise yards. V are concerned, the scathingly graphic descript Society for the Prevention of Cruelty to Ch

English high-rise flats applies equally well to those in Ulster: ' . . . usually small, inadequately fenced, unimaginative, grubby and, of course, outside . . . Playgrounds consisting of a few swings and a battered-looking roundabout, of a few concrete shapes (a cannon? a boat? a fortification? a large pipe—or was this left over from something else?), a playground with a 'grit' floor capable of scarring knees for life . . . others floored mainly with mud . . . '[13]

A fortification? The fortress image is virtually a reality now, with the presence of machine-gun posts on the roof, barbed wire, sand-bags and sentries in the street below—especially at Unity Flats, a ghastly, expensive blunder, a residence for several hundred Catholic families placed at the foot of the Shankill Road, the very gateway of Orangeland. Before decay and the Army moved in, it is not difficult to imagine the effect this new, glittering edifice must have had on the Shankill Protestant who had perforce to pass the Flats every day as he went between the shipyards and his smaller, dirtier home on the Shankill. Imagination can supply all the details; it is scarcely necessary to turn back to the Press reports of nightly jeering, stoning, flag-waving.

As in any working-class area, conformity is everything, but here, in the Protestants' districts, it takes the unique form of a fierce pro-Britishness, a terrible nationalism probably without parallel anywhere else in the world. The kerbstones are painted red, white and blue in threes, often the full length of the street,[14] and bunting may flutter overhead all year round, zig-zagging up the street from eave to eave.

At the street's end, the gable wall is a medium much explored by the political commentator and artist. The favourite drawing is, of course, that of William III on a white horse—sometimes beautifully and skilfully executed. There might be a lack of contemporary evidence that William's horse at the Battle of the Boyne on 12 July 1690 was actually a white one; still, it is conceivable that a folk hero should be depicted on a mount of any other colour.[15]

Slogans, equally fascinating, vary from the pejorative—'No Pope Here', 'Fenians out',[16] 'F— the Pope'—through the admonitory 'Remember 1690', 'Not an Inch'—to the lyrical—'We will forsake the Blue Skies of Freedom for the Grey Mists of an Irish Republic.' The latter, of uncertain origin, has enjoyed a long spell of popularity.

Every available space must be used for some relevant message. On a wall, gable or pillar box it is brief, succinct: 'Rem 1690', 'F.T.P.', but, by a skilled operant, a really long wall can be made to carry a fully-worded denunciation of the present Stormont regime (thought over-

liberal on the Shankill), to draw attention to the increasing influence of the Papacy, and to hint at radical solutions.

'To Trade with Ulster's Enemies is Treachery,' reads the heading of a widely-displayed notice. It is printed by the 'Ulster Constitution Defence Committee', and forbids purchase of butter and other goods made in the Republic of Ireland; it also urges refusal of Republican money (legal tender in Northern Ireland). This campaign has been at least partly successful; it is now well-known that to offer Irish coins in any shop in a Protestant area is to risk, at the least, a brusque refusal. Our milkman, on the surface a quiet, sane individual, once spat on an Irish shilling I gave him and threw it to the ground.

At one or two street corners there is posted a microcosm of the Ulster Protestant predicament. Two notices in careful juxtaposition read: 'Get right with God', 'Vote Unionist'. A Union Jack dominates the message,[17] and underneath is a picture of the current Unionist candidate for the constituency.

The Union Jack is *de rigueur* in Protestant areas, especially during July when the Orange celebrations take place and an exterior flagpole is almost standard equipment on each house. From 1971 onwards, however, there has been an interesting vogue for displaying the Ulster Flag as an alternative. (This flag consists of the Red Cross of St George on a white ground with, in the centre, the Red Hand of Ulster surmounted by a crown.) It is not easy, standing so close in time to events, to say how significant this change of symbol is; it may well indicate growing right-wing disenchantment with a Westminster government believed to be not unduly concerned with the plight of the Protestant majority. To suggest that this change in emphasis implies UDI aspirations[18] may be too rash a conclusion to draw, but, as an unusually observant English commentator has reported, ' . . . there is a stream of the Orange mind which muses still on revolutionary independence rather than provincial status'.[19]

It would also be erroneous to conclude, though it is a pleasant fantasy, that the flags, the bunting and the folk-art are all manifestations springing spontaneously from loyal bosoms. On the contrary, the group's need for a display of solidarity can lead to over-valuation of the symbols and thence to ugly incidents. A Cockney visitor, unaware of the necessity to display a flag, was bullied and intimidated into the acute ward of a mental hospital, broken in body and spirit.

Other images are more ambiguous—here a ladder, there a Latin cross or a five-pointed star, renderings vaguely Masonic, basically ornamental. In the Protestants' houses, pictures of the Royal family are still in evidence, but now you can also buy pictures of the Revd Ian Paisley, framed in red, white and blue, in varying sizes for public display or private contemplation, and these prints seem to be just as popular as those of the 'Royals'.

The identification of the Catholic population with the Republican cause is a much more uncertain quantity than that of the Protestants with Unionism, the latter virtually total in the Protestant working class. However, there has been a consistent failure among pro-British observers to estimate accurately the proportion of committed Republicans among Catholics. Some months ago Major-General Tuzo, the GOC, put this proportion at one-third. Even then the figure seemed over-conservative; it must appear an even more timid estimate now, when the British Army has been for so much longer a visible focus for old, cultural hatreds—now, too, that tales of looting, destruction and brutality have become current,[20] and now that the extremist organisations have more and more taken upon themselves the images of protectors, of freedom fighters. Even as early as February 1969, when the Civil Rights campaign was in its infancy, the Strathclyde University survey found that 66% of the total Catholic population did *not* support the Stormont regime.[21] And now there is no doubt that the introduction of internment, more than any other single factor, has swung Catholic opinion massively behind the Republican cause and, to a lesser extent, the terrorist organisations.

But, even yet, the Unionist-controlled *Belfast Newsletter* is capable of writing in a leading article: 'If there were to be a secret ballot among the Catholic population tomorrow for an answer to the question: "Which of the two Armies would you like to see pull out at once?" there would scarcely be one who would hesitate for a moment before writing the initials IRA.'[22] These opinions spring from a somehow touching belief that no one could possibly take up an anti-British stance unless he were intimidated into doing so. No man can be blamed for being intoxicated with patriotism, but this distressing condition is bound to form his prejudices and colour his opinions. There can be no doubt that, whatever the position may have been some years ago, when Catholic areas were indentifiable simply by the absence of Loyalist paraphernalia,[23] things are very different now, and Republican sympathies have become overt. The Irish flag is erected at windows and on barricades faster than the Army can remove it, and wall slogans, virtually unknown in Catholic areas in 1969, now jostle for elbow room. They sprawl across every wall and gable and even spill over on to the surfaces of the pavement and the road. Contrasting with those of the Protestant areas, they are manifestly the work of a youthful minority and are exclusively non-sectarian. Favourite slogans are: 'Join the IRA', 'Army bastards out', 'British out', 'Join Fianna Eireann', 'No tea for Dad's Army'. (The latter refers to the popular British TV series lampooning the wartime Home Guard.) But it is striking that there are never signs of aggression directed towards Protestants as such; the security forces, the British and Stormont governments bear the brunt. In some streets the nameplates have been removed and the streets re-named in favour of Irish patriots—this mainly in the New Lodge Road area, a stronghold of the Provisional IRA.

Inside Catholic houses the outward religious forms predominate—there are statuettes and holy pictures in every living-room—again in sharp contrast to Protestant houses, where the only hint of religious observance is likely to be a prayer-book in a corner or a bible on the shelf. In the Protestant areas—public religiosity, private politics; in the Catholic areas—the exact reverse.

While family size varies, roles within the family generally differ little between the Catholic and Protestant working-class communities. Nor do they differ much from those of Hoggart's general model of working-class parents. The stance of the husband is that of the breadwinner, the provider—overtly if not actually the dominant partner, somewhat exploitative. The tacit assumption is that certain duties are owing to him; he expects to be waited on and cooked for; domestic tasks are strictly out of his province. He may put up shelves or a cupboard, but will not wash dishes. He might take the older boys out to a football match, but will never take the younger children to the park. If he and his wife go out with the baby in the pram, *she* pushes it. He may grow roses, but never 'flowers'. He may devote his time to pigeons and budgerigars in quantity, but never in pairs or singly, in circumstances that would suggest that the birds were 'pets'. The standards of behaviour in a Belfast working-class area, Catholic or Protestant, are in fact as rigid as those of the Royal Enclosure at Ascot. The growing boy, picking his way painfully through the maze of selective reinforcement, slowly learns the hard lessons of life—the right language, the right clothes, the right degree of male assertiveness—that men are men, and boys are men. Undoubtedly, the sex roles in Northern Ireland are even more stereotyped than Hoggart described them in English cities, although the differences are largely in degree, as in most other 'settler' communities, such as the United States and Australia. This is to some extent an expression of population selection, the immigrant conforming to his stereotype—tough, male-dominant, hard-working, go-getting, self-starting. It also reflects the strong religious conservatism in both Protestant and Catholic camps, with its roots deep in Old Testament Judaism—a religion in which the male daily gives thanks to God that he was not born a woman.

Thus role-reversal, common now in both Belfast and Derry, where, very often, the husband is unemployed and the wife the breadwinner, can strike particularly deeply at the male psyche. Harold Jackson, in his penetrating study, *The Two Irelands*, refers to 'an embryo matriarchy in which the traditional dominance of the male has been steadily eroded'. He enlarges on this further:

'In a society which still observes the sort of cultural mores which persisted in England in the nineteenth century, this has had deep social and psychological effects. There is a constant need for the men to assert their masculinity.

'Often this takes the form of excessive drinking and gambling. Pilot studies of the effects of long-term unemployment have suggested that it eventually saps virility in the strictly physical sense and this, too, may well have set up considerable stresses. The result of it all has been a growing incidence of vandalism and blind destruction of public property—walls defaced, telephone kiosks destroyed—and a growing inclination to combat authority in the most flagrant way possible. This was one way to show that masculinity was still potent. Allied to the political grievances already simmering away it is evident why the riots that eventually broke out took the form they did.'[24]

This is not quite so evident to everyone. One is tempted to ask what form a riot might take other than that of physical violence. But Jackson still has, I think, made some valid points, and has established at least one additional observation point that will be useful later in this book.

As for the woman—even with an unemployed husband—looking after the children is still very much her job, and she is almost entirely responsible for their care and discipline. To hold down a full-time job and look after a family as well might seem an impossible task—and would be, but for the frequent existence of the 'extended' family. This is a central feature of Ulster working-class structure, and I believe that its importance has not been fully recognised. In the older streets, the population has been virtually static for four or five generations and there is almost invariably a grandmother or some other relative under the same roof or very close at hand to help with caring for the children during the day. A certain cooperation among neighbours, too, becomes essential as family size grows, and the young mother comes to rely heavily on these various sources of support. Given a high male unemployment rate and families of up to fifteen children, this arrangement can be very satisfactory. It has worked well in the past in the Ulster cities, as it has in other cultures where similar conditions obtain (for example, in the southern states of the USA), but at the same time it is clear that here, already, were some of the ingredients for the general breakdown of family structure and discipline that was to shock the world's Press in 1971. In short, these were rehousing of family units in high flats or housing estates entailing the break-up of the extended family, new environments where the children's play and other activities could no longer be supervised, large families, male unemployment, and, very often, an absent working mother in a cultural setting where the father could not modify his role so as to take over adequate care of the children. (This applies particularly in Derry, where there are plenty of jobs for women in synthetic fibre plants, but much fewer opportunities for men—so that the woman is very often the financial provider.)

It is true that many of these problems could apply, though perhaps not all together, in any large city; in themselves they do not necessarily constitute a formula for conflict or for the breakdown of the social fabric. What then of the factor unique to cities in Ulster—the religious problem?

The Catholic and Protestant Churches in Ireland are the most conservative in Europe. The Catholic Church, for example, still insists on totally segregated education; Catholic children may not go to State schools, but only to Catholic schools. The Catholic hierarchy have recently displayed a certain hypersensitivity in this area, to the point of objecting to the term 'segregation'. They admit only to having 'separate' schools; the term 'segregation', if used, is usually qualified by a phrase such as 'as Protestants call it', or even 'as extremists call it'.[25] However, when a group of children is separated from other groups for all lessons and activities purely on the grounds of their religion, when they cannot live together or play together, it is difficult (without, at present, going into the defensibility of the system) to feel that any other term applies.

To be fair, the Catholic Church's response can, to some extent, be read as over-reaction to a rigid Orange control of education that has been in operation for over fifty years. Many observers believe that this herding of children into different camps at a very early age is a major cause of the present conflict, but look in vain for any signs of willingness on the part of Church leaders even to discuss educational policy. The route by which each faction has reached its now deeply-entrenched position is traced in the next chapter. It is sufficient at present to emphasise that this segregation is total throughout Ulster—except in one or two small country schools; also, that the Catholic Church sets such store by this aspect of its mission that, as recently as 1969, parents who sent their children to a

State school were threatened with refusal of communion by their bishop. This process carries on inevitably into housing, where each community has grown up around its church and its school. Some mixed areas have persisted until recently, but housing has become more and more polarised over the past four years, in slow trickles following damage and intimidation, as well as in the massive population displacements of 1969 and 1971.[26, 27] While prior to this the Falls and Shankill areas were the only districts where there was heavy predominance of either Protestants or Catholics, the point has now virtually been reached where there are no mixed areas, where, in Belfast, Catholic and Protestant may never again live side by side.

The conflict in Ulster has entered its fifth year. Adults, and older children, still have memories of days outside the barricades when they could go freely to and from school, playgrounds and football matches, of nights when the pavements were lit and there was glass in the shop windows. But the real tragedy is that the youngsters who are now just old enough to run and play in the scarred and barricaded streets have never known anything different.

Given present indications, what can they look forward to? Environment, for a Belfast working-class child, means a decaying terrace house, a high flat or a concrete housing estate with less than one-third of the Government-recommended playing space. He will mix, play, and be educated exclusively with his own religious group, and may never see a child of the 'other' group, except across a barricade. His home area is divided from other areas by armed sentries and steel barriers; he cannot leave this area, and to go out even in his own street after dusk is to court injury or death. The street is in pitch darkness at night, carpeted with stones and broken glass, littered with burnt-out, rusted skeletons that, perhaps a year ago, were cars, buses and lorries. His home is overcrowded, his father has a high chance of being unemployed,[28] and poverty may be acute.[29] He is brought up with a fear and hatred of members of the other religion that will last him all his life.

I embarked on this chapter with the intention of disproving a current belief that Belfast is made up of ghettos; it didn't bear thinking about. I had hoped to show that the ethnic enclaves of Watts, Harlem, Chicago and North Africa were far worse, far more violent, squalid and terrifying. Now, having visited and studied them all, I know that the reverse is the case.

Ghetto, enclave, compound—these are ugly words. But when the evidence from Belfast has been reviewed, the conclusion must be that the ghetto metaphor has rarely been more aptly used.

3

Papers and Preachers

PREMIER IS PELTED BY REPUBLICAN MOB

Ulster's Prime Minister, Mr Brian Faulkner, escaped unscathed from a missile-throwing mob in East Belfast last night, as new violence flared on the Newtownards Road.

As he left Mountpottinger police station shortly before 10.30 p.m. teenage hooligans from the Roman Catholic Short Strand area, who had congregated on the opposite side of Mountpottinger Road, pitched pennies at Mr Faulkner.

Belfast Newsletter, 15 April 1971

CHURCH ATTACKERS ESCAPE CENSURE BY GOVERNMENT

As the Newtownards Road simmered last night following the previous night's Protestant extremist rioting there was a marked reluctance by members of the Government to condemn the petrol-bomb attack on St Matthew's Catholic Church by a crowd estimated at 2,000.

Mr Brian Faulkner, the Prime Minister, was in the area last night—and he ran into trouble, on the Protestant side. Stones and a bottle were thrown and the Premier was jeered and booed by a crowd of about 700 but they missed and he sped off in his car.

Irish News, 15 April 1971

Viewspapers reflect attitudes—as distinct from newspapers, which may challenge or modify them. In Ulster, a neat expedient avoids any unseemly struggle for control of the news media; there is, simply, a Unionist and a Republican version of each event.

An English journalist wrote of Belfast's two morning papers:

'. . . like the Chinese water torture, it's the regular, relentless drip that does it. Just such a drip is the daily appearance of the *Newsletter* and the *Irish News.*'[1]

Growing independence of a colonising Power, coupled with new feelings of national identity, is probably the setting *par excellence* in which perception can be distorted and information processed to the point of unrecognisability. Rhodesia is the best-known modern example—although totally blank columns on the front page suggest a degree of inexperience. By contrast, the Belfast morning papers have acquired, over the years, an ability to process news items that is probably unique. How have they used this ability in order to shape and mirror attitudes?

Gordon Allport, in *The Nature of Prejudice*, writes that two major processes are applied to items of information in an atmosphere of high prejudice.[2] He calls them *attenuation* and *assimilation*. Attenuation, literally 'shortening', means selection of items from the whole in order to convey a desired impression; assimilation means that the items are moulded and coloured on the base of a pre-formed set of perceptions. The Unionist-owned *Newsletter* and the Catholic-owned *Irish News* may well see the same event quite differently. In the following example—the brief saga of a provincial confrontation in two versions—the writers could probably escape charges of conscious distortion altogether.

Irish News, 5 July 1971

Fifteen arrests were made by police in Lisburn at the weekend when a Protestant extremist mob stormed Chapel Hill, the town's Catholic quarter, and petrol bombs were thrown and property damaged before they were driven from the area. And last night the committee of St Patrick's Parish, Lisburn, angrily repudiated Press reports that a flag was flown from parochial property or that any provocation was offered to the Protestant mob. It was emphasised that there was not a Catholic on the streets during the disturbance and the confrontations were between the Protestant mobs and the security forces.

Belfast Newsletter, 3 July 1971

A policeman was injured when a mob surged towards a Roman Catholic church hall in Lisburn early today after terrorists had planted firebombs in big stores in the town. Dozens of extra police including the special patrol group were rushed into the town when a 200-strong crowd gathered outside St Joseph's Roman Catholic Church hall where a dance was taking place. A tricolour had been waved from a window in the hall.

The more glaring discrepancies between the stories are not worth pointing out. But note, first, that while the *Irish News* reports a 'Protestant mob', the *Newsletter* attenuates this information to, simply, a 'mob'. The information on terrorist activity in the *Newsletter* is attenuated in the *Irish News* to the point of being omitted entirely; the same thing happens in the *Newsletter* to the Protestants' petrol bombs. *

The fire-bombs, assimilated by the *Newsletter*, become 'terrorist' fire-bombs. The crowd, for the *Irish News*, is a 'Protestant extremist mob'. The *Newsletter*'s phrase 'a policeman was injured' suggests an occurrence almost *sui generis*, and its account of the crowd's behaviour, the almost inevitable consequence of strong provocation.

Thus, almost effortlessly, does fact become propaganda without quite becoming fiction. Thus can a political or sectarian viewpoint be strongly pushed without recourse to more formal editorial comment, as in a leading article. The *Newsletter* can easily project its own emotional reaction in headlines like 'Anger Mounts over Killings' while, as the *Irish News* sees it, 'An explosion took place . . . ', an event scarcely attributable to any human agency.

It is quite possible, too, for an editorship to engage in vigorous sectarian sniping from behind the skirts of its own readership, this by judicious selection for the Letters to the Editor page. Views expressed, the heavy print reads, are not necessarily those of the editorship. But it must be seriously doubted whether the *Newsletter*, in particular, can so easily shift responsibility for the most depressing collection of sectariana in newspaper circulation. Catholics, says one correspondent, are 'breeding like rats'. Another, using a favourite term, refers to Catholic housing estates as 'rabbit warrens'. I once wrote to the Editor, suggesting mildly that Catholic children were human too, but my letter was neither published nor acknowledged.

The other recurrent theme in *Newsletter* readers' letters is 'if Catholics aren't satisfied they should go back to the Republic'. This is the logic of the Mad Hatter.[3] The great majority of Catholics, having been born in Ulster, cannot go 'back' to the Republic, or anywhere else.

But the prejudiced individual, anyway, attributes alien qualities to an entire group, generalising from a few. Prejudice means judging an individual largely on the basis of the *group* he is thought to belong to, rather than by his own qualities. So any Negro is dirty and lazy, a Scotsman tight-fisted and a Jew grasping. Prejudice is not, however, quite the same thing as bigotry, which takes the process a stage further to an unshakeable belief in the inherent superiority or 'rightness' of the group that the bigot belongs to. For example:

The *Protestant Telegraph* (circulation 25,000) organ of the ultra-

Protestant Right, deals generally with statements, in various forms, of this concept. South of the Border—this line conceived as all but the edge of a flat earth—there lie only the 'Paddy-fields'. North of the line, every prospect pleases, with the exception of the 'Fenians', 'apaches', 'Papists', or 'Irish negroes'. In Catholic areas 'one passes strange individuals who speak a weird language and look barely human'.[4] On Sunday they go to the 'mass-house' or the 'scene of the abominable Mass'.

But Eden can be purged. The title 'Shooting Vermin' heads a short section dealing with the use of police guns against rioters. The writer muses: '. . . 10-bores would be more effective. The big slug makes an entrance hole far bigger that the exit hole of an Armalite or any other projectile of similar calibre.'[5] Shortage of factual material does not embarrass the prejudiced assimilator—in this case the Revd Ian Paisley, Editor of the *Protestant Telegraph:* 'Still another Belfast Corporation Public Library has been opened . . . some 15,000 books are available for borrowers. One may conjecture that among this plethora of books are many which can be construed as being covert anti-Protestant and anti-Unionist propaganda, together with those that are the blatant balderdash eulogising every aspect of Popery and rebellion.'

And a useful 'Is It True' heading forms an umbrella wide enough to cover the most improbable of rumours: 'Did Cardinal Conway, heavily disguised, meet IRA gunmen?'[6]

It is difficult to know just how seriously much of this is intended. But Malcolm Muggeridge has observed how often reality defeats the best effort of the satirist. Charles Dickens, creator of the *Eatanswill Gazette,* would have felt something of the same artistic despair had he been able to read, of the Chairman of the Community Relations Commission: '. . . a hideous monster, a political frankenstein, a Papal ogre . . . From some Co Down necropolis they dug up this creature, charged him with undying bitterness and seething with hatred. His hypnotic stare, his featureless and unsmiling face terrorise his meagre audiences to fulfil his every satanic wish . . .'[7]; of Major Chichester-Clark (when Prime Minister): '. . . a harmless, aristocratic nincompoop . . . traitor . . . knave . . . despicable creature . . .'; of Harold Wilson: '. . . this melodramatic pomposity with the tiny, nasty, obnoxious little voice . . .'; of Brian Faulkner (when Prime Minister): . . . preposterous little puppet . . . ARCH-LIAR . . . little LIAR . . . leering lopsidedly, the obstinate little lackey of Westminster . . . absurd, fatuous, stupid little man . . . jelly baby . . .'[8]

This style is unlikely to influence the opinion of anyone who did not already share these views. In some ways it is rather splendid, especially since sales bring a handsome profit. But one of the more regrettable aspects of this type of publication (of which the *Protestant Telegraph* is only the most widely circulated example) is its implicit denial of adult or even human status to political opponents and to people who do not share the writers' religion. Very many young people and children see these publications or hear the terms relayed by their parents. They cannot appreciate any tongue-in-cheek aspect where it exists, and they are bound to reproduce aggressive language and behaviour which appears socially sanctioned.

It might be going too far to include emotive or bigoted journalism among the *causes* of a sick society. But, carrying misconception and fear from one member of the community to another and from adult to child, it can be one of the virulent organisms which ensure that the disease remains endemic.

Communication of a germ is most easy when the organism is rendered vulnerable; no one knows this better than Ian Paisley, founder of the Free Presbyterian Church, and Ulster's most skilful communicator of aggressive anti-Catholicism. On the platform, his techniques for probing and utilising anxiety are extremely subtle, and stem largely from his Evangelical background. He has been trained in the preaching style of John Wesley and his successors, one that depends on raising anxiety about the after-life to a level at which, as Wesley found, the individual becomes highly susceptible to suggestion and at which, therefore, conversion to a new religious faith can most easily occur. (This type of technique has also been used in modern psychiatry, in the form of treatment by 'abreaction'. The patient is encouraged to imagine himself in an event which he fears, perhaps with the help of drugs, and to express his anxiety in the form of uncontrolled emotion—crying, laughing, shouting. He is believed to be highly suggestible in the period immediately following this 'abreaction'—and so is amenable to suggestions that his fears are groundless, or to introduction of means to cope realistically with them.[9])

Paisley speaks to a special audience. The Evangelical approach, strong in Ulster, depends on clear and personalised concepts of good and evil—God and the Devil. He is without difficulty able to transform the evil, feared object, in a couple of easy stages, from a horned Lucifer to a terrifying amalgam of the IRA terrorist, Paddy the green simian, and the Scarlet Woman of Rome. Disaster is predicted. Heedlessness of his message will mean loss of Protestant privilege, of land and jobs—then a return to the terrors of the Smithfield fires and the tortures of the Spanish Inquisition. With anxiety raised, presumably, to its maximum, the hope of salvation is held out—a return to the old Protestant authority-figures, the upholders of the *status quo*, the RUC, the B-Specials, Stormont. This message speaks strongly to most Ulstermen. Not only that, it gains for Paisley himself a near-Messianic status . . .

'In Ulster's darkest hour, increasing numbers of disillusioned people turn towards the only man who, by God's grace, can lead this Province as it should be led. Dr Ian R. K. Paisley is a man of the people, a man of Ulster, a man of God.' *(Protestant Telegraph,* October 1971)

'I am not a prophet, but I may be the wife of one,' Mrs Paisley said recently. Prophet or no, the image, in the hands of this very clever strategist, is not entirely undeserved; Paisley is consummate master of the self-fulfilling prophecy.

Herein lies a mini-science in itself. I illustrate: if you meet a man on a foggy night and tell him he is about to walk over a cliff, this is a *self-defeating* prophecy. But if you had phrased it differently, telling him that since there was a cliff ahead he would stop or turn back, you would then have made a *self-fulfilling* prophecy (except in the unlikely event of his wanting to walk over the edge of the cliff).

Taking this into the political scene—if you want to brand the police as brutal, corrupt and biased, you will not do so effectively through an obscure paragraph in an obscure newspaper. Instead, having ensured maximum advance publicity, you assemble your followers near a police-station and, through a megaphone, shout your opinion until the police, goaded beyond endurance, appear with batons and tear-gas to illustrate it. A few hymns and a prayer having been included, you can later describe the event that was so brutally interrupted as a 'religious meeting'.

In the same way a government march ban is a bumper present in terms of publicity to Civil Rights protesters and other dissident groups. With the props in the shape of soldiers, guns and barricades supplied, the protesters have nothing to do but march. It is inevitable that the 'peacefulness' of the marchers and the 'brutality' of the security forces, who are under the ban bound to use physical force to stop them, will appear in sharp contrast. And it happens that violence done to a law is not so easily photographed as is violence to the person. The cameras whir, and the protester can go home well satisfied, his cause justified in the eyes of the viewing public.

Dogma leads inevitably to organisation.

'I have said before and I repeat today—the Orange Order is the backbone of Ulster.' Brian Faulkner, 12 July 1960.

To what kind of organisation, cap in hand, comes the man with ambitions to be Prime Minister?

All of Ulster's Premiers have, in fact, been members of the Order. One-third of all male Protestants are members, and membership is virtually a must for a prospective Unionist MP. The Order has been portrayed in various lights, from a Protestant Ku-Klux-Klan to an irrelevant nonsense compounded of sashes, tribal heroes, folk jingles and dancing in the streets. One thing is beyond doubt, however—that is, that the Order is the most powerful single political organisation and educative force in Ulster. Its origins explain why.

The word Orange, the name of a small principality in the Rhône valley, came to Ulster with William III in 1690, via a long line of Low Countries princes. But the Order itself is only 180 years old. It came into being during territorial disputes between Catholics and Protestants in the 1790s as a Protestant defence organisation, drawing its first members from a group calling itself 'The Peep O'Day Boys'. As the name suggests, they raided and terrorised Catholics' houses at dawn, usually on the pretext of searching for arms, and engaged in early battles with a militant Catholic group called 'The Defenders'. Freemasonry was strong in the area; members of the new, rapidly-growing Protestant society in Belfast and Derry, whose members were mainly Masons, adopted a similar system of password, signs, and local 'lodges'.

Several Masonic Lodges already bore the title 'Orange' and in these the victory of William III in Ireland had been toasted almost since 1690; the title 'Orange Society' seemed a natural choice, and the warrant for the first Orange Lodge was probably taken in 1795. The more militant members of the society succeeded in driving many Catholics from County Armagh and, having now a machine capable of making and consolidating territorial gains, they attracted support from Protestant land-owning gentry and clergy. The latter rationalised their membership by stressing its 'defence of the faith' aspect and in return offered the Order a new, sanctified respectability.

Basically, little has changed. The Order is still dominated by clergy who insist that its aims are primarily mystical. But in an official history by three Orange clergy, confusion becomes apparent as early as the foreword![10]
' . . . it is to this movement that the world owes the establishment of the concept of Civil and Religious Liberty . . . emphasis is laid on the maintenance of the Protestant Religion and the Act of Union.'

The book's dedication, quoting from a traditional Orange toast, is 'Tó the glorious, pious and immortal memory of King William III . . . ' The dots, the remainder of the toast, read:

' . . . who saved us from rogues and roguery, slaves and slavery, knaves and knavery, popes and popery, from brass monkeys and wooden shoes; and whoever denies this toast, may he be slammed, crammed and jammed into the muzzle of the great gun of Athlone, and the gun fired into the Pope's belly, and the Pope into the Devil's belly, and the Devil into Hell, and the door locked and the key in an Orangeman's pocket for ever, and may we never lack a brisk Protestant boy to kick the arse of a papist; and here's a fart for the Bishop of Cork.'

Again, on education: 'The Roman Catholic policy of strict educational apartheid has produced a ghetto mentality among her people . . . Nowhere is discrimination more apparent and its effects more obvious . . .'

But, in the same document: 'The 1923 Education Act caused an uproar among the Protestant people who saw in it a betrayal of . . . the opportunity of ensuring that their children were brought up as Protestants. They were told that there was no guarantee that Protestant teachers would teach in Protestant schools . . . '

'The 1947 Act was to put back the clock making it possible for anyone as well as a Protestant . . . to teach in a Protestant school . . . '

For teachers at least, the Order's priorities are clear enough. And for the mighty in his seat—provided he wanted to stay there—continued loyalty to Orange ideals has always been essential.

Lord Craigavon, Ulster's first Prime Minister, offered simply a 'Protestant Government for a Protestant people'. He told the Ulster Commons in 1934, 'I am an Orangeman first and a politician and a member of this Parliament afterwards.' Lord Brookeborough's assertion that he would never employ a Catholic was no secret. At an Orange demonstration in 1933 he said, 'I have not one about my place,' and, in 1934 'I recommend those people who are loyalists not to employ Roman Catholics.'

Terence O'Neill, deposed by an Orange 'parliament' because of a supposed soft line with Rome, said in his farewell speech: 'It is frightfully hard to explain to a Protestant that if you give Roman Catholics a good job and a good house they will live like Protestants . . . He cannot understand that if you treat Roman Catholics with due consideration they will live like Protestants in spite of the authoritative nature of their Church.'

An individual who fears loss of security may regress to a form of words or behaviour appropriate to a much earlier stage of human development. A community, its identity under threat, may also regress, perhaps to the defence mechanisms of a tribal society. The Orange Order provides, through its rituals, the avenue for both.

Harold Jackson, trying to probe the motivation behind Orange processions, draws attention to the 'importance of territory to all animals, including man, as a source of security'.[11] A procession defines boundaries (as between Catholic and Protestant areas, for example) and serves as a display of unifying symbols and the trophies of past warfare. All the trappings have antecedents as old as Man himself—the pictures of dead, charismatic heroes carried on poles, the sashes, the ceremonial swords, the night-long roar of the terrible Lambeg drums, and the huge fires where the children dance on the eleventh night of July, the eve of the anniversary of the Battle of the Boyne.

For these children, the learning of group identity has begun already; young boys join the Order at the age of eight. Under considerable pressure from their peers in Protestant schools to join, they find it almost impossible to leave without being branded 'Fenian-lovers' and victimised in various ways. Their major events are at Easter and in August, when they parade —or rather, they are paraded. One vivid eye-witness account from a teacher: ' . . . At 6.30 a.m. the streets leading off the Crumlin Road and opposite the [Catholic] Mater Hospital has each its quota of police and military units. From 8.30 a.m., the Junior Orange Parades began their ritual . . . As each band came level with the hospital the drums got an extra bang and the war dances became more vicious and meaningful . . . as a woman from the country, I was frightened. One wrong word or sign . . . The most tragic factor . . . was that the Junior marchers' ages ranged from 8 to 12 years and they looked like any other little boys, tired and a little puzzled; and in them I could detect no ill-feeling at all. The provocative actions came from . . . the hoary old masters of ceremonies in the 50-plus age group who were the obvious organisers.'[12]

Of course, in the same way that not all bigots are Orangemen, all Orangemen are not bigots. In many isolated areas the Lodge is the sole meeting-point for social and cultural activity. However, the Junior Lodge remains the major agency for the introduction to Protestant children of cross-cultural myths, fears and hatreds, and for the sanctioning of verbal and physical aggression against Catholics. Every Ulster Protestant child, for example, knows this song (from *Orange Loyalist Songs*, Belfast, 1971):

> I was born under the Union Jack
> I was born under the Union Jack
> Do you know where Hell is?
> Hell is up the Falls
> Kill all the Popeheads, and we'll guard Derry's walls.

> I was born under the Union Jack
> I was born under the Union Jack

Falls was made for burning
Taigs are made to kill
You've never seen a road like the Shankill.

I was born under the Union Jack
I was born under the Union Jack
If Taigs were made for killing
Then Blood is made to flow
You've never seen a place like the Sandy Row.

I was born under the Union Jack
I was born under the Union Jack
If guns were made for shooting,
Then skulls are made to crack
You've never seen a better Taig than with a

bullet in his back.
and:

My old man's an Orangeman
No Fenian can deny
He loves to wear the Orange sash
On the twelfth day of July
He looks a lovely picture
Marching with the rope
He'd love to march right on to Rome
And hang the cursed Pope.

(When sung, the word 'cursed' is not pronounced as printed. Taig and 'Fenian mean Catholic. The terms have, broadly, the same emotive quality as 'nigger'. The former of the songs was the subject of the first prosecution under the new Incitement to Hatred Act (NI), brought by the Community Relations Commission. The publisher, defended by a barrister who was also a Unionist MP, was acquitted.)

The (Protestant) Ulster Volunteer Force is a much more nebulous affair. Like the IRA, it is an illegal organisation, banned in 1966, taking its name from an earlier (legal) force formed by Edward Carson in 1912 to resist Home Rule. It does not at present have the status of a formal, coordinated movement, though its name appears in every gable in Protestant areas, and claims of wide potential support have been made. The fact is that the UVF is now in much the same position as was the IRA before 1970, little more than a legend projected from the 1920s, but with emotional roots in the population deep enough to provide an immediate rallying point should community security be threatened. This stage, for the UVF, is likely to come soon, probably in the event of transfer of further power to Westminster or withdrawal of British troops.

There has recently been a vogue, among Protestant gangs, for carving the initials UVF on the hands or arms of Catholic captives; also, with a metal punch, for using the initials to deface Irish coins. These youths argue that the organisation's obscurity implies efficient counter-intelligence rather than lack of support, and have showed me pubs and clubs where the leaders met. The Army believe that, although these claims may be exaggerated, a vast stock of legally-held Protestant arms, plus a quota of trained ex-B-Specials, would enable an efficient force to be mobilised in seventy-two hours or less. Much of the machinery, together with a chain of command, now exists in the form of organised vigilante groups who nightly barricade and patrol their own areas—an extremely ominous development. The Shankill Defence Association alone, for example, has some twenty thousand members, and the ultra-Right Ulster Vanguard at least ten times as many.

It is likely that, in the same way as Sinn Fein, the Ulster Vanguard movement will emerge as the political wing of an armed guerrilla group—in this case the UVF, or an equivalent. This latter could, for example, be the uniformed Ulster Defence Association, formed in 1972 and about ten thousand strong. Evidence available at the time of writing, however, makes it unlikely that the UDA is a front organisation for the UVF, which continues to have a separate existence.

Through these organisations, the next few years are almost bound to see a huge upsurge of organised Protestant violence.

The most outstanding story to come out of the present conflict in Ulster concerns the growth of the IRA, Oglaigh na Eireann, within only eighteen months, from an almost mythical entity to a membership and support unprecedented in its fifty-year history. Historically, however, the IRA stands first in the long line of modern guerrilla movements that extends from the 1916 Irish Rising, through the Chinese and Russian Revolutions, to the present Vietcong and Tupamaros campaigns. Since 1945 several urban guerrilla groups have, in fact, been IRA-trained. In 1955, plans to send Eoka recruits from Cyprus to Ireland for training in guerrilla warfare were only terminated by the London Agreements. IRA expertise was believed to have been responsible for the relative success of the Spanish Basque guerrilla movement, Euzkadi Ta Azkatasuna, in a series of bombings and bank robberies in 1968-9; now there are reports that IRA 'training officers' have been appointed by terrorist groups in Italy and the Dominican Republic, and by the American and Israeli Black Panthers. Further reports tell of a conference held by the Arab PFLP (People's Front for the Liberation of Palestine) and the Black September Movement in Dublin in May 1972 and, later in the same year, of strong IRA representation at a terrorists' summit meeting, probably in Tokyo.[13]

The Irish Republican Brotherhood became the IRA in 1921 in a bid to resist partition, effective under the new Government or Ireland Act (1920). Low-key terrorist activity was more or less continuous until 1962 when, after an abortive border campaign (Operation Harvest), the Republican Publicity Bureau announced: 'The Leadership of the Resistance Movement has ordered the termination of the campaign of resistance to British occupation. All arms and other materials have been dumped and all full-time active service volunteers have been withdrawn.'

The statement blamed public apathy for the decision, and the *New York Times* commented: 'The original IRA and Sinn Fein . . . have been condemned by the most deadly of all judgements, public indifference . . . '

Herein is the clue to the reason for the organisation's massive comeback in 1970-71. It is highly unlikely that any fully worked-out plan for a Northern campaign existed before then, but it must soon have become clear to militant Republicans that by provoking the Army to increasingly harsh measures against the Catholic population they could soon gain the wide community-base essential to all urban guerrilla groups. By the end of two years of arms searches and arrests, ending with the introduction of internment, the Belfast IRA battalions were composed entirely of local men. 'Officers' are full-time, with no other employment; they are financially supported by house-to-house collections, and to a lesser extent by proceeds of bank robberies by SAOR Eire. 'Volunteers' are in normal daytime jobs, operate in the evenings, and are recruited from the age of seventeen.

The split between the Provisionals and the Officials came into the open in 1969, but is the product of conflicting ideologies that have divided the theorists, at least, since partition. (Catholic ghetto children call the two groups the 'pin-heads' and the 'stickies', because of the different way they wear their lilies at the Easter parades.) The Official wing is traditional Marxist; it stresses political activity with the aim of establishing a socialist republic throughout Ireland. Violence is viewed as a means of defence or retaliation rather than as a major vehicle for political change.

The Provisionals, on the other hand, with the avowed aim of permanently toppling Stormont, have much closer ties with mainstream Irish nationalism. John Stephenson (alias Sean MacStiofain), head of the Provisional Army Council, proposes a return to the Gaelic Athletic Association's ban on soccer and a revival of the Irish language so as to give 'a cultural seal to Ireland's separateness'. In the hope of re-uniting Ireland with bomb and gun, the Provisionals' calculation is that of every terrorist organisation in a colonial situation—that they can, in the end, persuade the occupying Power that it is just too costly to hold on. There must certainly be comfort for the Provisionals in the results of a survey by National Opinion Polls, published in January 1972, which indicated that, for the first time, a majority of the British electorate favoured withdrawal of the British troops from Ulster.

But there is, at the same time, no comfort for the British Army or electorate in this apparent split in militant Republicanism. With the man in khaki on the streets a visible focus for old, cultural hatreds, the organisation retains massive working-class support. Given in addition (like the Vietcong) proximity to a sympathetic landmass, a guerrilla movement is in a prime position from which to strike. In Ulster, the point at which the use of increasingly harsh measures by the occupying regime becomes directly counterproductive has long since been reached and passed.[14]

What motives lead a young boy into the IRA or its junior branches? These will be explored in more detail later. In effect, added to the social pressures on an Ulster Catholic child (enumerated in Chapter 2) we now find a cultural hate-figure, introduced into a child community with learnt Nationalist ideals—the British soldier. The IRA has been diligent in transplanting these fears and aggressions into active support for its cadet organisations. Several other guerrilla groups, notably the NLF in Vietnam and EOKA, have succeeded in attracting and training pre-adolescent members, but the IRA has a machinery for this purpose second to none.

Fianna Eireann, the largest junior branch of the IRA, has an estimated membership of several thousand boys in the eleven to sixteen age-group. Their full uniform is similar to that of the Boy Scouts—green beret, green shirt and yellow scarf; most members just wear green berets, however. (*Bona fide* Scouts in Belfast now *don't* wear their berets when outside; it has become an intensely emotive article of dress. The sight of one could start a riot, and has.) The Fianna Cub section is of even more recent organisation; boys join between the ages of seven and eleven. This section is often in the charge of female leaders, assisted by 'Volunteers' from the Fianna.

The Fianna merges to some extent with the older Catholic Boy Scouts of Ireland; some CBSI groups remain based, as before, on traditional Scout principles. Others are undoubtedly front organisations for militant anti-

Army activity and fertile recruiting grounds for the Fianna and, later, the IRA.

This applies also to numerous youth clubs and billiard clubs. In one billiard club that I saw in the North Belfast area, the billiard tables are in a large front room; a small room at the back is used for group instruction and demonstration of guerrilla techniques on two nights a week. A complicated system of signs (which I did not learn) warns of the Army's approach. Attendance of boys in their early teens may reach a hundred; local women supply tea and buns.

There are reports that some guerrilla manuals, such as *Booby Traps, Total Resistance,* and *The Anarchist Cook Book* have reached Ulster from the United States. These do not seem much in evidence; there is probably little that the Belfast experts could learn or teach from books.

The sophistication of some of the guerrilla techniques taught to children has already been illustrated. The youngsters have graduated, with age and experience, from stone-throwing and bin-lid-banging to the calculated use of deadly weapons—petrol-bombs, nail-bombs, sub-machine-guns. (Members of the Fianna receive training in the use of fire-arms in Dundalk and Donegal.) Younger children learn, among other things, how to make incendiary devices by tying a bundle of matches half-way down a smouldering cigarette. This amounts to a five-minute fuse. Before the Army started searching children and their schoolbags in 1972, youngsters used to transport gelignite around, sometimes in blazer pockets, or strapped between shirt and skin. (This is not quite as dangerous as it sounds; without a detonator, gelignite does not explode easily.) But its use in bombs is a different matter . . . 'Youngsters and older children are ideal material for the work of planting bombs and rigging booby-traps,' reads a directive to IRA members. 'They attract less attention and suspicion than adults, are more sensitive to rewards, and ask no questions. If captured by the British Army or security officers they are unable to provide information about their employer . . . More gelignite nail-bombs and petrol-bombs must be readily available. British Army patrols can be lured into ambushes more easily when children, youngsters and women are the bait . . . '

Like the expertise, the gelignite is handed down the line. Although some gelignite is undoubtedly smuggled from Eire, approximately three million pounds is annually imported legally from the Republic and Scotland for use in quarrying, road development and housing. Security on most sites is inadequate, and much of the imported gelignite has been proved impossible to trace. But, while it may have been manufactured in Irish factories, the common phrase 'explosives from the Republic' implies a good deal too much.

In passing the explosives to children to be used as nail-bombs, or to be lit and thrown directly at troops, adults depend on the Army's hesitating to open fire on youngsters. Placed to the rear with more deadly weapons, they use the children as decoys to draw the troops into range by throwing stones and petrol- and nail-bombs. Children, with limited death-concepts, unable through immaturity to anticipate all the risks of their actions, have accepted this role without hesitation, and in other settings besides Ulster.

Perhaps the situation closest to that of Belfast and Derry was that in the *communes* of Paris and Lyon a century ago; children, building barricades and throwing missiles at the National Guard, defended 'no-go areas' in both cities for several weeks. Writers, including Victor Hugo, have celebrated the 'incredible bravery' of these children. They would have done better to deplore their exploitation—as might General Grivas, who writes admiringly about terrorists' use of children in Ireland (Chapter 8). But this is a common feature of guerrilla warfare—although the terrible Simba tactic of using young boys as a human shield (p. 126) was little more than an ill-planned act of desperation. In Vietnam, it is probably the old hierarchical structure within the family that prevents the majority of young children from taking anything more than a very minor role in guerrilla warfare —carrying messages, for example (the ideal task for children brought up to be inconspicuous). Even these youngsters tend to be isolates. They have drifted citywards from bombed or plundered villages; the father may have disappeared, and the mother is over-extended. Often they attach themselves to 'uncles' in the NLF, which organisation becomes to some extent a surrogate family for them. Back in the villages, youngsters are confined to shouting at American soldiers (when the village is hostile) the one English phrase they all know: 'Fuck you . . . !'

In Belfast too, only a part of the youngsters' activity is organised by adults. Most of these juvenile gangs form spontaneously, adopting the uniform and tactics of their elders as naturally as did Edwardian schoolboys under the influence of *Scouting for Boys*. The patrol, keystone of Baden-Powell's system—the group of five or six boys under two natural leaders—is an almost universal pattern. My most vivid memory of Derry's Bogside is of watching half a dozen small boys, under the direction of a teenager, throwing stone after stone at a target marked on a wall, then at the chalked figure of a soldier. When it was dusk they met another group, then went off round Aggro Corner and down one of the tiny, winding, incredibly steep alleys that lead into the heart of the ghetto; I didn't follow.

But this is Derry 1973. The story of the rebirth of terrorism in Ireland has carried us, chronologically, well beyond Derry 1968, the year of the first major riots in the city—when the IRA scarcely existed. For completeness, I conclude with a short factual account of the sequence of events that led up to the explosive month of August 1969.[15]

Demands for Catholic equality first gathered strength in the west of Ulster, where there was felt to be a particularly heavy bias in the allocation of jobs and housing by the Unionist Councils. In 1964 the Campaign for Social Justice was formed in Dungannon and later, in Belfast in 1967, the Northern Ireland Civil Rights Association was formed, modelled on the National Council for Civil Liberties. Its aim was 'to assist in maintenance of civil liberties, including freedom of speech, propaganda and assembly'. It would 'advance measures for the recovery and enlargement of such liberties and . . . take steps as the Association deem[ed] necessary to that end'.

It sought, in particular, the following social reforms: an end to discrimination in housing and employment by Unionists, new electoral boundaries that would allow fair representation to Catholic electors, universal adult franchise (i.e. without a property qualification), and abolition of the company, or multiple, vote. The last two objects became crystallised in the slogan 'one man, one vote'.

Although the Association was non-sectarian, it was as inevitable that it should obtain strong Catholic support as that it attracted immediate and fierce Unionist opposition. Its earliest public demonstrations—a sit-in protest against a housing allocation at Caledon and a march in Dungannon—provoked the first of a series of militant Paisleyite attempts to counter the marchers with physical force. On 5 October 1968 the Minister of Home Affairs, William Craig, banned a Civil Rights march in Derry; enforcement of the ban by the police with repeated baton charges and heavy use of water-cannons brought Northern Ireland the widest Press and television coverage in its history.

Further worldwide sympathy with the Civil Rights movement came after the notorious Burntollet ambush, when some five hundred student marchers were the victims of a planned attack by a huge Protestant mob armed with stones and nail-studded clubs. Two outstanding factors emerged in subsequent reports: connivance in the attack by local Unionist leaders (including a Justice of the Peace), and apparent police apathy in the face of repeated onslaughts on the marchers. Anti-Stormont feeling was raised to white-hot intensity, and there were further huge demonstrations in Newry and Derry. Then, in March 1969, a Government Commission under the chairmanship of Lord Cameron was appointed to hold an enquiry into 'the causes and nature of the civil disturbances in Northern Ireland on and since 5 October 1968'. Their main conclusions on causes were:

'A rising sense of continued injustice and grievance among large sections of the Catholic population . . . in respect of inadequacy of housing provision by certain local authorities . . . [and] unfair allocation of houses built.

'Complaints, now well documented in fact, of discrimination in the making of local government appointments.

Complaints, again well documented . . . of deliberate manipulation of local government electoral boundaries . . . in order to achieve and maintain Unionist control of local authorities and so to deny the Catholics influence in local government proportionate to their numbers.

'Fears and apprehensions among Protestants of a threat to Unionist domination and control of Government by increase of Catholic population and powers . . .

'Early infiltration of the Civil Rights Association . . . by subversive left-wing and revolutionary elements.

' . . . deliberate and organised interventions by followers of Major Bunting and the Revd Ian Paisley [which] substantially increased the risk of violent disorder . . .'

The Commission also found that 'the police handling of the demonstration in Londonderry was in certain respects ill-coordinated and inept . . . Available police forces did not provide adequate protection to marchers at Burntollet Bridge . . . and in Londonderry on 4 January 1969.'

It was probably, again, a deep sense of injustice that brought violent reaction from Derry's Catholic population in July 1969, when the Stormont Government allowed a massive Apprentice Boys' demonstration to follow its traditional route through the city. (The Apprentice Boys of Derry, a branch of the Orange Order, march annually to celebrate the closing of the city's gates in 1688 against the army of James II by thirteen apprentices. Membership is adult; the Apprentice Boys are not to be confused with Junior Orangemen.)

The day began badly, with white-gloved Orangemen shouting abuse from the city's north-west wall and tossing pennies down into the Bogside slums. But worse was to come; from jeers and a few missiles as the procession passed Waterloo Place (the edge of the Catholic area) events escalated to a pattern that is now familiar—barricades, overturned vehicles, stone-throwing. Children, in relays, kept teenagers on the roof of Rossville Flats supplied with petrol-bombs well into the night. The police, making repeated sorties into the Bogside, fired a total of a thousand cartridges and fourteen grenades of CS gas, but were constantly forced to retreat under a furious hail of bombs and broken paving-stones. (As it is technically a 'smoke', CS gas can be used without breach of the Geneva Convention.) On the following two days the police were joined by the B-Specials and Protestant crowds; it was only the last-ditch expedient of bringing in British troops, who established and guarded a boundary between rioters and police, that prevented the first deaths occurring there and then.

But meanwhile in Belfast each group, Protestant and Catholic, had identified strongly with its counterpart in Derry. Some jeering from Catholic residents of Unity Flats, together with a rumour of stones thrown at a Junior Orange Parade, was enough to mobilise a furious Shankill crowd. On the night of 14 August mobs surged down the narrow streets that separated the Falls and Shankill areas and, by morning, entire rows of houses had been burned down, some four thousand people were homeless and six names, including that of Patrick Rooney, aged 8, headed a death roll that was to reach hundreds within two years. The Army arrived in strength; pessimists feared that they might have to stay until Christmas.

4

The Cost of Commotion: 1969 Onwards

How have the riots affected the mental health of Belfast people in general? Although very much worse was to come in the next two years, the upheaval of August 1969 still had immense 'shock value', with rioting and population displacement on a scale unprecedented since 1886. By the end of September, there had been 328 injuries requiring hospital treatment, 8 deaths and 113 arrests.[1] At least 500 houses had been damaged beyond repair and 4,000 people rendered homeless. Some 200 streets were heavily barricaded and guarded; much of the city was having its first taste of what was an effective martial law.

One event had followed another with startling rapidity; in the working-class areas the old familiar life of the street had gone, it seemed, for good. The reek of CS gas and burnt timber hung in the air, the once ideological barrier between Protestant and Catholic had become an ugly actuality, and tension in the tiny streets was electric. Piles of sandbags at corners, tortured coils of barbed wire on the pavement, all became less evident as the Sappers moved in, as sentry-posts acquired, subtly, the trappings of permanence—armoured glass, corrugated roofs. Social life in the community contracted disastrously as, nightly, each street was sealed off from the next; with the inevitable loss of most transport services and entertainments, rioting or 'area defence' became the sole *raison d'être* for adolescent groupings. Rumour was rife and suspicion of strangers almost pathological.

By the end of August there seemed to exist, at least on the surface, a sort of weary adaptation. But an English journalist wrote: 'A serious wave of mental illness has developed in Belfast as a result of last week's rioting and continued tension. Doctors are reporting influxes of patients with mental

45

breakdowns. In some cases, the symptoms are so severe that they have had to be admitted to mental hospitals.

'One doctor in the Shankill Road area—the scene of one of Belfast's most bitter fights—said he had prescribed more tranquillising drugs in the past five days than he usually does in a year. He had seen cases where men and women were unable to stop weeping . . . Even doctors in relatively peaceful areas have more people who have either developed new nervous troubles during the riots, or whose previous mental disorders have been considerably aggravated.

'It is feared that, so long as the street barricades stay up, and as long as homelessness and lack of domestic stability persist, the incidence of mental illness will not improve, and may even grow worse.

'A spokesman for Purdysburn Mental Hospital estimated that, since the riots, admissions had increased by 25%. People who had been burned out of their houses and who had been taken into the city's makeshift refugee centres had, in some cases, completely broken down. They were chiefly suffering from anxiety states and reactive depression. In two cases men had been admitted who had become obsessed by the fear that they were going to be shot . . . '[2]

A general practitioner in the Falls Road was reported as saying that tranquilliser prescriptions had gone up tenfold, but that even that kind of treatment was no longer sufficient as a palliative as the street fights and explosions continued. He said: 'You used to be able to cheer people up, tell them there was nothing wrong with them, give them a tonic and send them away happy. Now they are breaking down and weeping in the surgery.'[3]

A Londonderry GP, in a letter to the Press, said: 'I would like to make an appeal . . . to those people who are engaged in "political persuasion", operating in Creggan and Bogside. In the last 24 hours I have been called to help some of their victims . . . One of these victims has since tried to commit suicide and another has required urgent treatment for an anxiety state . . . a large number of decent people are so frightened that they depend on tranquillising drugs to keep them going. The long-term effect of this is almost too serious to think about.'[4]

Another Londonderry GP agreed. He is reported as saying: 'Hard-liners threaten moderates because they aren't extreme enough . . . Only yesterday I saw a letter to a patient saying: "Your name is on the next bullet. Get out now while you have a chance." There's been a huge increase in the number of people suffering from the tension. Many people are so frightened they just can't go on without tranquillising drugs.'

This doctor said that the number of people needing tranquillisers had increased five-fold in the past year. 'I know of one girl who went out with a soldier. It was enough to start people whispering threats about coffins, wreaths and hearses as she walked in the street . . . And I know of a woman who was frightened out of her home because she helped a soldier escape from a burning jeep during a riot.'[5]

More recently, a GP from the same area gave it as his opinion that the prolonged experience of sporadic bombings and intimidation was proving even more stressful to the people of Derry than that of street rioting and destruction.

He wrote: ' . . . during the last few weeks a terrible increase of mental distress has been evident. Daily I have a long procession of people coming to my surgery with tales of fear and of shattering experiences. I am having to prescribe increasing amounts of sedatives, and for the first time in my career I find it necessary to issue tranquillisers in substantial amounts for children, in order to get them a night's sleep, and to help them through school.

'For the first time I have begun to fear that the people of Derry will not be able to survive the trouble without a complete breakdown of society as we know it.'[6]

A consultant psychiatrist said that, from August 1969, there had been a sharp increase in the number of attempted suicides in the province. The majority of the attempts were by women, and he believed that this increase was connected with the recent disturbances.[7]

At the same time, other observers were quick to discount the Press reports as grossly exaggerated. The Secretary of Belfast's main mental hospital said that there had been no significant increase in admission rates as a result of the riots.[8] In Londonderry's mental hospital, the opinion of consultants was that disturbances in the city had not been responsible for any important change in the number or pattern of admissions. My own impression, formed while working in this hospital during a brief period in 1971, was that very few admissions could be directly attributed to riot conditions *per se*. The most common problem, in fact, was rather that of elderly patients who could normally have been adequately cared for by their families, but who had to be admitted because the added burden of continued street disturbances and damage to property meant that the responsibility of an ageing relative had to be, at least temporarily, shed. Others, chronically inadequate patients, clearly were using the pretext of riot disturbance to support demands for admission. This group would have sought a sheltered environment in any case, riots or no—they simply had new support for their claim to illness.

A consultant psychiatrist working in Belfast, after a careful survey, concluded that the existing psychiatric facilities were adequate to cope with the situation, and that the appropriate use of tranquillisers by the family doctor provided effective control, quickly relieving people who developed fear and anxiety symptoms.[9]

Thus far, no definitive picture emerges. The Press reports are certainly disturbing, but they are also highly anecdotal, and do not lead to any conclusions that can be generally applied. The two reports that appear most objective also appear to be contradictory.

In late 1969, I embarked on a detailed analysis of recent psychiatric admission figures, out-patient attendances and drug prescription rates. As well as trying to clarify the effects of the recent stress in Belfast itself, I hoped the results might help towards an understanding of the wider relationship between mental illness and the stress, direct and indirect, of open hostilities.

It is one of the more curious aspects of psychiatry that wars do not seem to cause or even worsen mental illness; very often the exact reverse is the case. In countries at war the admission rates to mental hospitals generally fall, though the reasons for this have never been generally agreed upon. Results of most studies have been ambiguous—just as, now, are the earlier reports from Belfast. Before going on to examine the results of my own study, I think that it is well worth while for me to include a brief review of investigations carried out in generally comparable conditions. Most of this work was done in wartime and, although there are important differences between wars and riots, the results of this earlier work are necessary to. an understanding of the figures from the Belfast study. When the two groups of results are compared, I believe that some very important conclusions emerge.

The civilian is the Cinderella of wartime, at least as far as psychiatric interest is concerned. There is a huge literature dealing with military psychiatry, but the effects on civilians are much more scantily documented.

Some useful accounts date from the Franco-Prussian War, when, in 1871, Legrand du Saulle reported a fall in admission rates to Paris asylums, and a general decrease in psychiatric out-patient attendances.[10] During the First World War, psychiatrists in London, Edinburgh and Belfast recorded a similar decrease in admission rates. An Edinburgh psychiatrist, indeed, believed that the effect of the war on the civilian population was minimal; the only people developing symptoms were those with a previous history of mental illness, who would probably have broken down anyway.[11]

The Spanish Civil War, a struggle more akin to the Ulster crisis than that of the First World War, seemed to provoke only a 'normal anxiety' reaction in Spanish civilians, as reported in the *British Medical Journal* of 1939. There was no increase in mental hospital admissions.[12]

During the Second World War crude figures for admissions to English mental hospitals showed a moderate decline,[13] although in some provincial towns, notably Bristol and Birmingham, there were slight to marked rises in the incidence of short-term neurotic reactions.[14] A striking fact was the lack of relation between severity of bombing and injury, and neurotic illness. A Welsh GP said that he had seen no neurotic reactions in areas where there had been severe bombing; all the anxiety states he had come across had been in people who had not been bombed at all. He called their condition 'anticipation neurosis'. Other British psychiatrists described a surprising lack of psychiatric symptoms among the injured or dispossessed. 'Having tasted the bitterness of homes gone and injuries sustained [they] had a serenity and calm which might be due to the fact that, having experienced their worst fears and come through, they were content to await the future and not rush to it—no evidence here of neurosis, either hysterical or depressive.'[15]

In general, London air-raids appeared directly responsible for relatively few cases of well-defined neurotic illness, although acute emotional reactions were fairly common.[16,17] In Coventry, also heavily bombed, there was a marked decrease in attendance at out-patient clinics.[18] In both cities, many chronic psychoneurotics showed a remarkable lack of susceptibility to acute reactions, and some appeared actually to improve.[19,20]

But the psychiatrists were to be challenged on their own ground. Dr T. Harrisson, who had a long-standing commitment with evacuees, wrote to the *British Medical Journal* suggesting that doctors found so few cases of neurotic illness in wartime simply because they looked in all the wrong places—that is, in the big cities, in towns, in air-raid shelters. People who had developed neurotic illness, he argued, had long since fled these areas; the act of flight was in itself often a pathological reaction. He detailed numerous instances of patients who had moved to quiet towns and villages—the most common clinical picture among this group being a form of chronic depression. He believed that the most disturbed cases came from 'nearly-bombed areas', and that uncertainty had been the most potent factor in precipitating their illness.[21]

A well-known psychiatrist, Dr Clifford Allen, offered in reply the theory that this depressive reaction was akin to the 'sham death' avoidance response seen in insects and animals under stress.[22]

In Norway during the Second World War there was a general decrease in first admissions to mental hospitals, but a marked increase in the incidence of acute psychotic reactions in males.[23]

In a major survey of 1966, F.C. Dohan examined war-time admission rates for schizophrenia in Finland, Sweden, Canada, Norway, Switzerland and the USA. He found that these decreased markedly in Finland, Norway and Sweden, and slightly in Canada and Switzerland, but that they increased sharply in the USA, especially among males.[24]

There has been a paucity of reports from Belfast, although severe rioting is nothing new to the city. There were series of riots in 1857, 1864, 1872, 1886, 1898, and 1920–22. There was minor rioting in 1964, and a prolonged but sporadic IRA bombing campaign in the 1950s, mainly confined to the border between Northern Ireland and the Irish Republic. In 1864, a Belfast physician reported: ' . . . great alarm and anxiety in nervous and delicate people produced a loss of sleep, strength and appetite, which, in many cases, terminated in low forms of disease.'[25] A Belfast consultant psychiatrist, Dr A. Lyons, whose study I have already referred to, found that the most common reaction was one of 'normal anxiety', and that there was a surprising lack of psychiatric illness directly attributable to the civil disturbance.[26]

The apparent effects of wars and riots in children are fully covered in Chapters 5 and 6, but it is worth noting this briefly: the general consensus of opinion among war-time child psychiatrists was that in children in blitzed areas and concentration camps acute neurotic reactions were very common, but almost always short-lived. In general, symptoms were more common in reception centres than in areas under direct attack.[27, 28]

Belfast, for the purpose of my study, was divided into three areas, graded for degree of disturbance. This grading, of course, was crucial; to make the study at all worth while expert assessments were necessary. Here I was very fortunate in having full cooperation and advice from the Army and the Police Security Department. The three areas were finally defined as follows.

Area 1, the most disturbed part of Belfast, represented the Vehicle Control Area (popularly known as the 'Curfew Area'). Here there had been fierce street battles and widespread destruction of property. Here, too, the Army had moved in until, by the end of September 1969, there was a massive military presence of some five thousand troops. Most of the smaller streets were barricaded, and all vehicle movements were restricted—hence the name.

Area 2 had a rather broader definition. The criteria were, basically, all *visible* signs of tension—barricades, boarded-up windows, reported incidents. There was a general expectation of trouble, but there had been no major experience of violence. This area, in spite of what might seem to be rather diffuse terms of reference, proved surprisingly easy to define, and shaded away quite abruptly into Area 3, the remainder of Belfast, where life was apparently going on much as usual.

Admission rates and out-patient referral rates to mental hospitals, psychiatric units and day hospitals for August and September 1968 and August and September 1969 were obtainable from central records, classified according to the major psychiatric diagnoses. These were coded for the address at which each patient lived, and so the totals could be grouped to correspond with Areas 1, 2 and 3.

In the first analysis, these figures were subdivided for age-group, religion and psychiatric diagnosis, as well as for sex and home area. But most of the resulting totals were too small for the use of statistical tests, so age-groups and religion-groups were combined. In any case, from inspection of the figures it did not appear that people of any single religion or age-group had been particularly vulnerable.

Next, tests of statistical significance were applied to all differences between the 1968 and 1969 figures.[29] The results of these tests are shown in Table I. (A statistical note: it should be emphasised that, for the sake of clarity, the table shows the totals only of the sets of figures to which the χ^2 tests were applied. Numbers from Area 3 were too small for statistical analysis, but no differences were apparent between the 1968 and 1969 figures.)

TABLE I

Admission rates and out-patient referral rates from Belfast areas during August and September 1968, and August and September 1969

Area	1				2				3			
Sex	Male		Female		Male		Female		Male		Female	
Diagnosis	Psychoses (290–299)[1]											
Year treated	1968	1969	1968	1969	1968	1969	1968	1969	1968	1969	1968	1969
Out-patient referrals	5	3	5	8	4	19[2]	8	7	2	—	—	—
Admissions	25	23	46	39	36	50	61	50	3	4	3	7
Total	30	26	51	47	40	69	69	57	5	4	3	7
Diagnosis	Neuroses (300–309)[1]											
Year treated	1968	1969	1968	1969	1968	1969	1968	1969	1968	1969	1968	1969
Out-patient referrals	13	20	19	20	14	26[3]	29	28	2	2	1	2
Admissions	53	50	41	36	48	58	56	78[4]	4	8	6	8
Total	66	70	60	56	62	84	85	106	6	10	7	10

Notes:
[1]W.H.O. International Classification of Diseases, 3-digit Categories, 8th Revision.
[2]1968/69 increase, all psychoses: $\chi^2 = 9.78$, $p < 0.01$.
[3]1968/69 increase, all neuroses: $\chi^2 = 3.60$ (for $p = 0.05$, $\chi^2 = 3.84$).
[4]1968/69 increase, all neuroses: $\chi^2 = 3.61$ (for $p = 0.05$, $\chi^2 = 3.84$).

Table I shows that there was a statistically significant increase in male psychotics referred as out-patients from Area 2, the area of intermediate disturbance.[30] Also for male psychotics there was a considerable rise in admission rates from the same area. However, for male psychotics from Area 1, the riot areas, these rates remained constant. As regards neurotic illness, there was in Area 2 a marked increase in the out-patient referral rate for males and in the admission rate for females, the differences in each case just approaching, though not quite reaching, the 5% level of significance. (But it should be pointed out that these returns are often several months late; in either of these groups one further 1969 referral would produce a difference at this level.) The noteworthy point is that these differences, again, occurred only in Area 2. There were no significant differences by year in admission and out-patient referral rates from Area 1.

Drug prescription figures for hypnotics, tranquillisers and anti-depressants in Belfast practices were analysed for the same periods.[31] these totals were grouped similarly, into Areas 1, 2 and 3. The boundaries of practices are not, of course, clear-cut, nor do they coincide with those of the three areas exactly, so that these areas, though comparable, are not identical for both parts of the study. Some practices were, as it happens, so large that they could not be definitely placed in one area; others had more than one surgery address, and a few had undergone changes of medical personnel between the two periods concerned in the study. The figures from these practices were omitted.

Results are shown in Table II. The only important change from 1968 to 1969 was in the prescription rate for tranquillisers, and this is isolated in Figure 1. This appears to show a fairly uniform increase across the whole of Belfast—but this has to be re-examined against a moderate population shift (see Table II) and the general year-to-year increase in the demand for tranquillisers in the community anyway. From the figures under Table II it should be clear that, with outward movement of the population to Area 3, mainly suburban, the difference in Area 1 does grow somewhat in importance—but there is still an across-the-board increase.

Has the increased need for tranquillisers perhaps got nothing to do with the riots? At first it seemed that this was so, especially when the corresponding rates from some provincial towns were analysed. These towns were widely separated and some distance from Belfast. All, too, had remained free from civil disturbance. The results of this analysis were surprising, almost disconcerting. The increases, three of which are shown in Table IV, ranged from 2% in a quiet farming community, to a dizzy 41% in a booming

TABLE II
Prescription rates for hypnotics, tranquillisers and anti-depressants in Belfast during August and September 1968 and August and September 1969

Area		1		2		3	
Year		1968	1969	1968	1969	1968	1969
Drug	Month						
Hypnotics: Prescriptions	Aug.	2,551	2,452	6,825	6,817	5,039	5,082
	Sept.	2,719	2,662	6,944	7,403	4,999	5,434
Total*	Aug.	187	166	446	439	369	364
	Sept.	173	177	436	492	343	478
Tranquillisers: Prescriptions	Aug.	2,244	3,035	5,594	6,804	3,979	5,236
	Sept.	2,315	3,101	6,118	7,942	4,238	5,820
Total*	Aug.	155	221	414	524	316	417
	Sept.	157	244	452	603	330	481
Anti-depressants: Prescriptions	Aug.	322	334	860	885	714	767
	Sept.	422	407	954	1,046	951	1,017
Total*	Aug.	26	32	73	89	62	84
	Sept.	34	41	76	106	73	99
Populations*†		83,673	86,523	205,066	223,441	152,403	170,746

* Thousands of units (tablets, capsules, etc.).
† *Population increase: Area 1:* 3·4 per cent. *Area 2:* 8·9 per cent.
Area 3: 12·0 per cent.

industrial complex. These figures in themselves would be the subject of a worthwhile study; it is enough at present to say that, in view of these large increases in provincial towns, and since the total demand for tranquillisers throughout Northern Ireland seems to increase annually by 10% *anyway*, there is no useful conclusion that can be drawn from this part of the study by itself.

TABLE III
Prescription rates for tranquillisers in six Belfast practices

Area		1				2				3			
Practice		A		B		C		D		E		F	
Year		1968	1969	1968	1969	1968	1969	1968	1969	1968	1969	1968	1969
Hypnotics: Prescriptions	Aug.	72	64	65	55	70	53	94	65	17	9	45	37
	Sept.	59	66	77	56	69	83	101	79	20	17	47	39
Total	Aug.	6,854	5,310	3,790	3,260	6,151	3,812	4,423	2,949	1,384	900	2,883	2,850
	Sept.	5,300	5,768	5,461	2,482	5,462	8,946	4,312	3,982	1,986	1,736	3,250	2,870
Tranquillisers:* Prescriptions	Aug.	49	84	38	48	42	33	78	77	11	8	61	61
	Sept.	46	72	29	68	36	45	95	102	15	5	84	83
Total	Aug.	4,984	7,890	1,450	1,786	3,434	3,160	4,347	4,934	908	680	4,465	6,425
	Sept.	4,184	6,476	1,068	2,306	2,856	3,924	5,156	6,150	1,376	450	7,017	6,140

* For 1968/69 increase in Area I, both practices: $\chi^2 = 60·85$, $p < 0·001$.
For Areas 2 and 3 combined, practices C–F: $\chi^2 = 1·92$, N.S.

But Table III and Figure I are much more striking in their implications. So far, only very large, diffuse, and partially overlapping groups had been studied, and it seemed likely that more marked differences were being concealed, or that any trends revealed were not uniform. So the next step was to select, from each of the three Areas, the two medical practices most centrally and typically placed—where the practice population, in each case, was most likely to be drawn exclusively from within that area. (None of these practice populations had varied by more than 5% during the yearly interval we are considering.)

TABLE IV
Numbers of prescriptions for tranquillisers in provincial towns

Town	Aug. 1968	Aug. 1969	% increase	Sept. 1968	Sept. 1969	% increase
X	627	757	20·7	672	802	19·2
Y	378	533	40·9	470	618	31·5
Z	195	200	2·6	212	203	—
N. Ireland	38,148	42,908	12·2	40,372	48,137	18·9
	Aug. 1967	Aug. 1968	% increase*	Sept. 1967	Sept. 1968	% increase*
N. Ireland	38,962	38,148	—	40,455	40,372	—

* But increase for all of 1968 on 1967 = 9·34 per cent.

These results are shown in Figure I and Table III. There are huge increases in the tranquilliser prescription rates in the practices in Area I as compared with little or no changes in Areas 2 and 3 and, statistically, this difference is highly significant.

FIGURE I

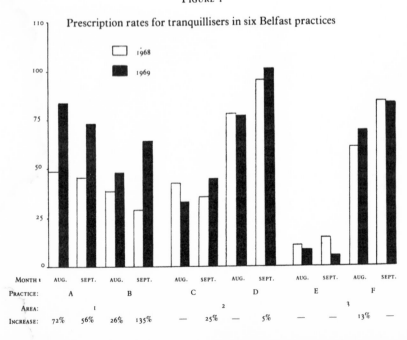

Prescription rates for tranquillisers in six Belfast practices

□ 1968
■ 1969

MONTH	AUG.	SEPT.	AUG.	SEPT.	AUG.	SEPT.	AUG.	SEPT.	AUG.	SEPT.	AUG.	SEPT.
PRACTICE:	A		B		C		D		E		F	
AREA:	I				2				3			
INCREASE:	72%	56%	26%	135%	—	25%	—	5%	—	—	13%	—

'The people were completely hysterical. I have never seen anything like it, even in the war.'

Comparison with war-time conditions came naturally to an English businessman, helping to evacuate office workers from a building wrecked by an explosion.[32] To anyone over thirty, it is the natural yardstick. But how valid a comparison is it? There are, in fact, some very important differences between the *modus vivendi* of the city at war and that of Belfast from late 1969 onwards. War-time writers, explaining the usual decrease in psychiatric admissions from most sectors, stress as reasons the strengthening of community ties and lowering of social barriers that a war brings in its train, as well as the vastly increased opportunities for employment, either in the armed forces or in home-guard type duties. Social isolation, usually a potent factor in the precipitation of mental illness, becomes a remote possibility in the general atmosphere of banding together in the face of a common enemy, in the united sense of purpose that war engenders.

The British 'war-time spirit' is now, of course, legendary. But in Belfast, at the time of this study, the community was deeply divided. The feared enemy was not so readily identifiable, and friendly and hostile areas were ill-defined. In some districts, the security forces were able to separate Catholic and Protestant homes by physical barriers, but in mixed streets this was impossible. In these areas tension and anxiety were continuously high; the threat lacked the episodic nature of series of air-raids, since bombers, gunmen and arsonists could, and still do, strike at any time or place. There were no cues such as air-raid sirens to allow at least a measure of physical and psychological preparation; trouble could break out at any moment, and neighbours regarded each other with the deepest of suspicion and fear.

For all these reasons it might be expected that effects on mental health would have been more deleterious than those of a war. The results of the present survey certainly seem to support this.

In the first place, riots, unlike wars, do not seem to *benefit* any kind of mental illness; Table I shows that there was no significant decline in any group or area. Neither have there been any reports of individual patients who improved as a result of any change in their conditions due to the riots. On the contrary, there was a clear deterioration in the neurotic group—and here a striking trend is apparent, in that the increases in admission and referral rates are seen only in Area 2, the area where trouble was expected, while the corresponding rates in the *most* disturbed area (Area 1) remain constant.

This trend may seem a little surprising, but it can now be viewed along with the findings of Lewis and Burbury, who reported an increased war-time referral rate for neurotics in provincial towns (as opposed to London), in reception centres, and from among people not actively participating in civil defence. The population of Area 2 is directly comparable with these groups, since among this group also there was no active combat or direct threat to life, but all the signs that such occurrences were daily anticipated. This area also included the major resettlement centres, such as the Beechmount and Andersonstown districts.

These changes in admission and referral rates are real changes, unrelated to population movement. In other words the new referrals were not simply people who had been living in the riot areas at the time of the first outbreak of trouble and had then been evacuated to the suburbs. If these were included in the admission figures from Area 2, any conclusions drawn from an apparent increase would, of course, be very misleading. But all the patients classified here under Area 2 had been living in this relatively undisturbed area throughout the trouble and thus had not experienced rioting at first hand.

Another similarity with the war-time findings is that the rise in neurotic illness occurs mainly in the female population. This sex difference is most likely, in the same way, an expression of different degrees of involvement. Men, with a more active role either in actual street fighting (a small group), or as vigilantes, are less likely to suffer from social isolation. The female, in her more passive role, has less opportunity to act on her anxiety and, as in the Blitz, might well be more prone to nervous symptoms. Culturally, too, she bears the major responsibility for the children.

It is even more instructive to compare changes in the admission rates for psychotic illnesses with those reported during the Second World War. In both Norway and the United States during the war years, admission rates for male schizophrenics and manic-depressives rose sharply. Norway, although occupied and labouring under massive intimidation, was not actively engaged in warfare; the United States also differed from the other countries studied in not being under direct attack.

Area 2 in this present study, though of course on a much smaller scale, forms an interesting parallel with these two countries, both in its comparative freedom from actual combat, and in the finding of a highly significant rise in referral rates for psychotic males.

Schizophrenics (who form the major group among psychotics) are believed to be particularly vulnerable to stressful life events. There has been a recent resurgence of interest in this aspect of the illness.[33,34,35]

So a distinct pattern now emerges from the admission and referral figures, taken as a whole. In riots, as in wars, the most damaging effect on mental health seems to occur in the fringe areas, where people are worrying about what might happen; not where people have actually experienced their worst fears, and where an active role, a feeling of *doing* something about it, has to some extent served as a defence against anxiety. While the activities of many people in the riot areas might be generally regarded as undesirable—that of expulsion or terrorisation of the opposite religious group—the fact still remains that this role appears to be therapeutic or preventive.

Stress is not an easy word to define, but the conclusion seems inevitable that stress productive of psychiatric illness is maximal in areas under *threat* of attack, rather than where there is active combat or direct risk to life and property. This is the Invasion Complex. Anxiety, not experience, is the outstanding stressor.

The prescription figures for tranquillisers are more difficult to interpret. In Figure 2 the increase in prescription rates appears large at first glance —but the percentage increase becomes insignificant when it is set against population changes, as well as against the increase over the whole of Northern Ireland and, more specifically, in the provincial towns.

FIGURE 2

Prescription rates for tranquillisers (all Belfast)

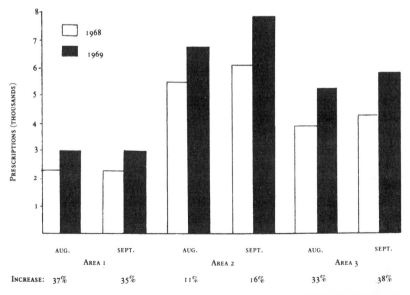

(To digress slightly—these latter results are rather intriguing in themselves, especially those from town Y, Coleraine, an industrial complex on the North Coast. Coleraine is favourably placed at the mouth of a broad river, and has been developing with immense rapidity in the past few years. It is also the home of a new university. It is hardly possible to attribute any increase in the use of tranquillisers there to events in Belfast or Londonderry; rather, on present evidence at least, it hints at some possible effects of urban living—perhaps on a population that has recently migrated from rural districts—and indicates the need for further study.)

So I have drawn no firm conclusion from the *total* drug figures. Also, in Area 3, an increase in the total numbers of anti-depressant tablets without a corresponding increase in the number of prescriptions could be little more than a function of the difficulties involved in evening travel between home and doctor's surgery. General practitioners seem to have been writing scripts for larger numbers of tablets at a visit in order to minimise the need for travelling to and fro, a reasonable, practical measure. But all these data would need to be viewed on a much longer time-scale for any pattern to emerge.

The study of practices A to F was much more rewarding. In each of these practices, the population is drawn almost exclusively from a single one of the three Areas. Table III and Figure 1 show a huge increase in the prescription rate for tranquillisers in Area 1, the riot area, as opposed to that in less disturbed areas. This almost certainly reflects a high incidence of acute emotional reactions in this area, not severe enough to require hospital treatment, but responding to drug therapy by the GP. These reactions were described by numerous observers at the time. Typical symptoms, such as dazed, stuporose reactions, or tremors and uncontrollable weeping, were very similar to those already described in both military and civilian practice. These symptoms represent a kind of letting-off steam, a release of pent-up emotional energy—and are common during, or immediately following, overwhelming stress.

To summarise, two kinds of psychiatric ill-health were evident following the outbreak of rioting in Belfast in 1969. There were, first, acute emotional reactions seen in people who had been directly exposed to riot conditions. These reactions normally quickly resolved either when the trouble had ceased or moved elsewhere, or in response to mild sedation. Secondly, there was psychiatric illness of a more incapacitating type which, for adequate treatment, required referral of the patent to a psychiatrist or his admission to a mental hospital. This second group, in marked contrast to the first, showed a significant increase only in areas *adjacent* to those affected by rioting.

This result is in line with those of most war-time surveys, and the general conclusion is that, in time of war or riots, any resulting increase in the incidence of mental illness is most likely to be seen in areas where people expect an outbreak of trouble, rather than in places where it has actually occurred (the Invasion Complex).

Although acute emotional reactions in riot areas can be dramatic, they are almost always short-lived; however, a superficial observer can easily be misled into overestimating their seriousness, and hence into pessimism about the effects on mental health of riots generally. In this situation more than any other, mental illness may be more apparent than real. The fact is that the spectacle of people weeping and trembling in the streets, or even an increased demand for tranquillisers, cannot be equated with a 'serious wave of mental illness'. This distinction is a vital one, since, in a stressful situation, a degree of anxiety is a normal physiological reaction; adrenalin is released, muscles are tensed, reactions are speeded, the organism is prepared for appropriate evasive or defensive action. When, at last, the tension is off, there is almost inevitably a certain 'over-spill'—the weeping, shaking reactions are then most prominent.

Lyons similarly found no increase in mental illness in the riot areas themselves, and reported that the commonest emotional disturbance was 'normal anxiety' reaction. When someone is said to be suffering from 'shock', it usually means that they are experiencing 'normal anxiety' in an acute form.

5

Ten Children

When there is a riot in your street, there are only two choices: you can either opt out or join in. It is true that you can, if fortunate enough, move out of the area altogether, but this is an option open only to a priviliged few. So basically, just like their elders, children respond to civil disturbance in either of two main ways. The aggressive reaction, where children join into groups to throw stones and petrol-bombs, is probably the most common, and certainly the most publicised. Perhaps the behaviour of these children is grossly disordered, but only a few of them reach the office of a child psychiatrist. When they do, it is usually because of some associated problem, such as truancy or school refusal, rather than for treatment of the aggressive behaviour in itself. In many areas, indeed, this activity is socially approved, openly by companions and tacitly by parents and relatives. In sharp contrast to the maximum publicity accorded this group, the huge number of children who suffer from serious emotional reactions have often been forgotten by everyone except their parents.

This section deals primarily with the types of emotional illnesses that have been seen in children in riot areas—symptoms that are clearly linked with the stress of street disturbances. Because—and this fact must be emphasised now—these psychiatric disorders have been much more serious and widespread in children than in adults.

Again, these psychiatric manifestations fall into two clear groups—acute and chronic. By far the most common are the *acute* emotional reactions experienced by a large proportion of the children who live in any riot area —symptoms that diminish and disappear shortly after the street has become quiet. The exact proportion is impossible to give in figures. However, from all accounts—parents, teachers, clergy and children themselves—I would suggest that scarcely any child living in riot conditions has escaped having at least some symptoms of acute anxiety—sleep disturbances, separation fears, school refusal, loss of appetite, bowel, gastric and urinary upsets, headaches.

This group of symptoms is akin to the 'normal anxiety' reaction seen in adults—with the difference that in the child population it was almost universal. How fair is it to apply a universal label of 'psychiatrically disturbed' to these children? Appallingly traumatic experiences are recounted by children in the case histories that follow; these were, of course, shared by *all* other children in the street. In the face of these, the children's intense anxiety should surprise nobody—nor should its manifestations, in the ways appropriate to their youth.

Although a child's acute distress is not to be minimised, I want to focus largely on an even more worrying problem: on those children in whom the civil disturbances precipitate psychiatric illness which persists, even worsens, *after* the period of acute stress—that is, after the violence has shifted to another part of town. These illnesses have often been incapacitating; many have, at the time of writing, been present for two or more years. These may be classed as *chronic* reactions.

Ten case histories follow. In the next chapter, these children's symptoms will be placed in a wider perspective, when the international literature on childhood reaction to stress is reviewed and compared with the reports from Northern Ireland.

While specific aspects of the cases may differ, the patterns of reaction are strikingly similar. This similarity in itself tells us a great deal about prevention and treatment—as well as providing a fascinating insight into the mental mechanisms by which a child deals with major stress.

All these children were referred to the Belfast Child Guidance Clinic by paediatricians.

Margaret is thirteen, and came to us because of fainting fits and symptoms of gross anxiety.

Her family had to move house during the riots of August 1969, when most of the property around them was destroyed by fire. Although their own home was not seriously damaged they suffered from intimidation and looting, and lost most of their belongings. After a short stay in temporary accommodation they obtained a small house in a street off the Crumlin Road—a street that was shortly to become one of Belfast's worst trouble spots. Rioting broke out in the street only a few weeks after they moved in;

for several nights in succession the area was a battle ground between crowds and troops, and the combatants frequently spilled into their front garden.

The first time that this happened, Margaret screamed, fell, lost consciousness and had to be taken to hospital—where all physical investigations were negative. She was discharged but, although her home district quietened down, she continued to have fainting fits both at home and at school. She ate and slept poorly, and began to lose weight. It was at this stage that we first saw her.

Margaret is the second of seven children, all living at home. Her father is chronically unemployed; both parents admit to having been very 'tense and edgy' since the outbreak of trouble. Her early history was uneventful. She had always been a shy, timid child, although difficulties in relations with other children now became much more prominent. On psychological testing, she turned out to be of somewhat below-average intelligence (Wechsler Intelligence Scale for Children, Full Scale IQ:71); all attainments were poor.

She was very nervous on interview, but was able to give a fair account of the initial traumatic events and the onset of her symptoms. Since August 1969 she had been kept awake on most nights by fantasies of fires and burning. There were numerous other fears, especially of loud noises, shouts and crowds. She was afraid to go out to school, and her school progress had suffered badly.

The fainting attacks continued for several months, sometimes after mild emotional upsets at home, but more often following experiences like seeing a crowd (an innocent one), hearing loud noises, voices raised in anger, or hearing references by her parents to the local political situation. Her parents, for example, told me about one fainting attack when she heard a quarry explosion, and another when they mentioned, in her hearing, a recent shooting incident.

Margaret says that she cannot remember any of these episodes, but she is now able to accept their relation to anxiety; though verbally limited, she has expressed many of her fears through drawing and play and has responded to explanation and reassurance. She has been free from fainting attacks for some months now; she has been placed in a class for slow learners, and is generally much less anxious both at home and at school.

Ten-year-old Marie was referred to the Clinic because of epileptic fits that seemed to be precipitated by attacks of intense anxiety. The youngest of four children. she lives in Belfast's notorious 'no-man's-land'—the maze of tiny, interlocking streets between the Falls and Shankill Roads.

Marie's father disappeared without trace eight months before we saw her; it is believed that there were political reasons for his departure. Her mother, herself an epileptic until the age of 14, has suffered from recurrent depression since the 1969 riots and complains of insomnia, tremors and weeping fits.

Marie has suffered from both major and minor epilepsy since the age of four. Her electro-encephalogram is unstable on hyperventilation, or over-breathing, but the test shows no focal abnormality such as might suggest early brain damage. Until August 1969 her epilepsy was adequately controlled with drugs (phenobarbitone and phenytoin), and she was having only three or four fits annually—but in August, following serious trouble in her street, the frequency of fits increased sharply and she had to be admitted to hospital.

She is of approximately average intelligence (IQ:90). Her mother said that she had always been a nervous child, who reacted to stressful events with tears and violent tantrums. Marie herself, a small, rather obese girl, gave me an animated account of her symptoms; she told me about their onset on a night in August 1969 when, she said, a crowd of Special Constables came up her street shooting at rioters, and when a factory nearby was set on fire. Her mother sent her upstairs, but forgot to pull down the blind. Marie sat at the window and stared out as if 'paralysed'. She said that she was not frightened at the time: 'It was just like the pictures.'

She described subsequent events vividly. 'A couple of days later we were going to school. We were late and running. Then I thought of *them* all running in the street and my legs went weak and I fell. I woke up in hospital and the policeman told me I'd had a fit.

'One day Mummy said, "Light the gas", and I thought of the CS gas and began to gasp and then fell down again. I had to be taken to hospital with another fit.

'The next day in school we had our Sound Book—you know, about the different sounds things make. The teacher said, "What goes ting-a-ling?" I thought of the fire-engine and I felt my hands shaking. Then I woke up on the floor.'

Marie continued to be severely handicapped by fits, mainly of major type, and generally preceded by over-breathing. As well as those she described, predisposing stimuli included bells ringing, the sight of an electric fire, seeing a crowd of boys running out of school, and an occasion when a friend said, 'I'm as hot as a fire.'

Marie was re-admitted to hospital and shortly later transferred to a child psychiatric in-patient unit. It was found necessary to shield her from local news programmes on television and also from references to the current situation, since the pattern of anxiety–over-breathing–fit has now become clear. On one occasion a fit was precipitated by a toy belonging to another child—a toy tank that fired sparks from a gun.

In therapy, an early response to signs of anxiety, together with consistent support, proved the most useful approach. In other words—it was possible to recognise and respond to fears expressed through play and conversation and so prevent anxiety rising to the point where hyperventilation could provoke a fit. The number of fits rapidly decreased; she was soon able to be discharged and, with community support for the family, has been at home for a year without relapse.

A child's electro-encephalogram, or 'brain-wave' pattern, sometimes becomes irregular when the child over-breathes; this can in certain conditions precipitate an epileptic fit. As over-breathing is often a component of intense anxiety, the rational treatment is reduction of this anxiety, either by removal of the threat, by psychotherapy, or by drug treatment.

Other physical ailments are classically linked with anxiety—as, for example, in the case of Sean.

Sean was just over eleven when he first came to the Clinic. During the few months preceding admission he had become increasingly irritable and tearful; he was also having frequent bouts of bronchial asthma.

He is the fifth of eight children, all living at home; he has always been timid and inclined to nervous symptoms. Otherwise his early history was uneventful. He is of above average intelligence (IQ:111).

Sean's mother said that he had changed considerably since a night in August 1969 when their street was the scene of a fierce gun-battle. He had developed intense fears of separation from his parents, was complaining of stomach pains, and had become more and more incapacitated by attacks of asthma. He had previously been inclined to be 'chesty', but this had never amounted to more than a slight cough; the first acute asthmatic episode was on this night.

Sean, a small, pallid boy with dark rings round his eyes, remembered the night clearly: 'I saw the crowds coming up the street. Somebody said: "They're beating up all the children; they're going to shoot them." I heard the guns and I felt sick and I shook all over and came out in a sweat. Then I started to wheeze and I couldn't breathe.'

After the local trouble had subsided Sean's symptoms continued unchanged. He lay awake at night for hours, expressing fears about fires and burning and also anxiety for his parents' safety if they happened to go out. Asthmatic attacks were sometimes precipitated by exercise, but the full cluster of symptoms (sweating, feelings of weakness, dyspnoea and broncho-spasm) was only seen following stimuli such as shouts, crowds, loud car engines and, once, a quarry explosion. They would also follow any reference by his parents to their own anxiety about the troubled local situation.

His mother said: 'I went into his room one morning and said, "I hear that a garage down the road has been blown up"—and right away he went terri-bly pale and his asthma came on. I shouldn't have said anything, but I had to tell someone.'

In spite of psychotherapy and drug treatment, Sean continued to be very limited in his activities for several months, and an improvement was seen only when his parents bought a new house in a quiet suburb (typical of Area 3). From the time that the deal was completed, even before the family moved, Sean's asthmatic attacks decreased in frequency, and he has now been discharged as symptom-free.

Janet, aged twelve, was referred to us as suffering from sleep disturbance, irritability and loss of appetite, her symptoms dating from shortly after an outbreak of street violence in her area. She is the eldest of five children; her mother said she had always been very 'highly-strung'. She has a long history of minor complaints and her school attendance has been poor. Her intelli-gence is 'low average'.

In August 1969 Janet's family were burned out of their home in one of Belfast's worst trouble spots. Since then they have squatted in various unoc-cupied premises. When seen at the Clinic, they were living in grossly over-crowded conditions, and twice had had occasion to move because of rioting and CS gas. The parents had found it extremely difficult to adjust to the situation; the father in particular, a quiet, introverted man, had been anx-ious and depressed for several months.

Janet's mother complained that she had 'gone to a shadow', that the slightest noise frightened her, that she would go nowhere by herself, and would not eat or sleep. Janet herself, a pale, undersized, timorous child, said: 'The Protestants came to our street and shouted "Get out", and the Specials were shooting at us. Then they burned us out. Now I can't sleep at night; I think about fires and burning and have to keep getting up to look out of the window. Any loud noise makes me cry and shake.'

Her symptoms—tremors, feelings of bodily weakness and tearfulness—continued from August 1969 until she was seen at the Clinic, several months later. Although she had not for the past three months had any direct exposure to rioting, she still experienced the symptoms in response to sounds like cars back-firing, quarry blasting, or shouts in the street. Her activities had become more and more limited; she would not go out alone, and fantasies of burning houses kept her awake on most nights.

When she was interviewed, on any reference to rioting or related occurrences, her fists would clench, her teeth close tightly together, and there would be noticeable facial pallor, sweating, and hand and jaw tremor.

Treatment was not very rewarding for her at first; although Janet clearly benefited from the use of Clinic media to express her anxiety to a responsive therapist, she was still living in surroundings that constantly precipitated her symptoms. There were drunken brawls nightly, rubbish was being burned all the time, and the sound of breaking glass was so common as to arouse little comment in the locality.

A Clinic Social Worker approached the Local Housing Committee, who were able to offer her family accommodation in a quieter suburb. Janet's symptoms rapidly cleared up from the time that this decision was known and, six months later, she was completely free from her previous handicaps.

Janet and Sean were two of the more fortunate children. Now with re-housing at a premium, other children are less so.

Anne was aged eleven when she came to us with a six-week history of hallucinations both at home and at school, and had upset her school friends and her teachers by screaming in the classroom, saying that she saw a 'bad man', and running out. She had been waking her parents at night by crying out in her sleep.

She is the second child of three in a Catholic family living in the disturbed Lower Falls area. Both parents admit to being very nervous and 'dreadful worriers'. Anne's early history was uneventful, although she was always a fearful, clinging child who reacted violently to feared objects. Once when she was a baby a bee had settled on her pram; she had screamed, turned blue, and fainted. Since then, her mother said, she had been terrified of anything that crawled or flew.

She is of average intelligence; her relations with other children were always good and her school progress satisfactory until she developed her present symptoms. These had become evident in late 1969 shortly after rioting in her street, when the family had to move house temporarily.

Anne is small, tense, and demonstrative. 'The Protestants came up to our street and let off bombs and guns,' she said. 'Then they ran away and the Army attacked our street. It was terrible; they were shooting and I saw a man all covered in blood. There were twelve of us children together in a house; we had to soak hankies in vinegar and put them over our faces because of the CS gas. We were very frightened; we were all crying.'

Shortly after Anne had the first of her hallucinations. She said that the figure she saw was a tall man with a big hat, brightly-coloured coat and frightening eyes. He varied in size from being only a couple of inches high to being 'ten times as big as a house'. He was, she said, a Protestant, because he was 'evil' and trying to kill her. Also, he was somehow 'different from us'. Sometimes he whispered to her and told her to do 'bad things'.

When Anne was first seen, this figure was appearing several times daily. There was no evidence of physical illness that might have given rise to such an hallucination (such as certain forms of epilepsy), and there were no signs suggestive of a schizophrenic process. The hallucination was precipitated by emotional scenes, by being out in the street, or by being in a crowd or seeing one. She had become terrified of going outside at all, and was refusing to go to school.

Anne attended for five interviews only. On the second, the hallucination was interpreted as a projection of her own fears. She had no difficulty in accepting this and in expressing her anxiety verbally, and the hallucination disappeared with dramatic suddenness. It seemed that she had previously felt inhibited in displaying anxiety directly about the local situation because she believed her mother to be particularly emotionally vulnerable. She remained free from perceptual disorders over the period covered by the next three interviews, although her mother reported some temper tantrums and handling problems. Further appointments were failed.

One could tentatively conclude that Anne's behaviour disorders, although most likely still expressions of high anxiety, proved more acceptable than did the earlier, more bizarre, symptoms.

Hallucinations in children are not so rare as might be imagined. An intense fear or wish not uncommonly takes visible or audible form—as described by Lauretta Bender. She also recounts how in some children, whose environments have necessitated a severe struggle to adjust to reality, personality integration is weak; part of the child's personality may thus be split-off, or 'projected'. Anne seems to fall into this group. In psychoanalytic terms, her hallucination was a 'projected id'. The man was always, she said, telling her to do 'bad things', and getting her into trouble. More simply, he was her 'bad self'—the one part of her personality she had not been able to integrate into the whole.

Anne is not the first person to have made this dissociation; it has always been a valuable release from anxiety. 'It wasn't my fault; the devil told me to do it,' says a child. The 'devil' of the Victorian Sunday-school is just such a projected figure—a scapegoat, the culture's collective id. He has appeared before now, too, to people under acute mental stress; Martin Luther even threw an inkwell at him. Luther never told us whether *his* devil had its classical form, with horns and a tail, but his post-mediaeval, monastic upbringing might well have invested it with just such a guise. Anne, in *her* culture, found that the devil was a Protestant. Why? A product of upbringing too?

The next three patients form a fairly homogeneous group. All are children of limited intelligence with long-term problems.

Peter is twelve. His IQ is 71, and he is an only child. His father has a long history of psychiatric illness; he is, at the time of writing, under psychiatric treatment, having developed symptoms of anxiety—mainly tremors and weakness when going out in the street alone—after the outbreak of rioting in their district. He is chronically unemployed.

Peter's most prominent symptoms also date from this period, although his parents say that he had always been nervous and inclined to headaches and spells of poor appetite. On the first night of the local disorders he complained of stomach pains and a headache, then fainted. Detailed physical investigations were negative, but he continued to complain of abdominal pain, fainting fits and, later, double vision. These symptoms were particularly related to fears of crowds and of being out in the street; when we first saw him he was quite incapacitated and had not been outside for several weeks.

His father noted that the symptoms became more prominent when he (the father) referred to the street disturbances, but he is preoccupied with the subject and is unable to prevent himself from doing so.

Peter was unable to relate any of his symptoms to anxiety and insisted that they had a physical basis. He did not, in fact, admit to any specific fears and little progress was at first made in treatment. But it was soon found that he could be steered, fairly easily, into what was overtly a discussion of his parents' and friends' fears about the street disturbances. In reality, the anxiety that accompanied this discussion on Peter's part strongly suggested that the fears were his own, so calm discussion and reassurance were directed towards the specific fears he described. The fiction that Peter had introduced was maintained throughout the period of treatment; I am not sure just how conscious of this curious piece of projection he was—perhaps not at all. But he has now become much less preoccupied with the local disturbances, and his symptoms have diminished to the point where he is again able to go back to school and to play in the street.

Maeve, aged nine, is of below average intelligence. She is the youngest of four children, and has a long history of school absences. She has always been a timid, clinging child, with a tendency to anxiety and tearfulness. Her street was disturbed by rioting in 1969 when her father, an active participant, was shot in the leg. Since then, she has been having frequent 'blackouts' precipitated by loud noises in the street, by being in a crowd, or by seeing one. Hospital investigations were negative, but she had been increasingly limited in her activities. When first seen, she was afraid to go out and hardly ever did so. Maeve improved greatly with psychotherapy—somewhat similar in principle to that given to Peter. She could not at first tolerate verbal discussion of her specific fears, since this provoked intense anxiety. But she has a certain facility for drawing and for writing poems (of a kind) that made it possible for her to communicate her fears indirectly. Thus was formed a 'bridge' to the stage where she was able to discuss her fears much more calmly and realistically and, even more important, where she was accessible to reassurance and explanation. The last stage in treatment was to facilitate communication at this level with her parents. We had several joint interviews which included a psychiatric social worker, but their success was limited by the parents' own anxiety.

However, Maeve herself is now much less anxious, and has had no blackouts for several months.

Aged ten, Emily has been an epileptic, of major type, from the age of six months. She was referred to us because of sleep and behavioural disturbances, as well as increasing frequency of fits. She is at a school for the educationally subnormal. She has always been a timid child and a poor mixer. She is the third child of four, all of whom live at home.

Rioting broke out in her street in late 1969; Emily's mother has suffered from headaches and nervous tension since. Emily started to have nightmares and to express constant (groundless) fears that her parents would be arrested and shot. She was afraid to go outside. Being in the street or in a crowd produced marked anxiety and over-breathing, usually followed by a fit.

The family has been housed in the suburbs; she is now much better, and fits are less frequent.

The remaining two children are current problems, and are typical of most being seen at present at the Clinic. The Ulster civil disobedience campaign in particular has struck very deeply at the mental health of a community where, in most Catholic areas, parents have suddenly found themselves in the middle of a tug-of-war between local vigilantes and rent collectors. With the city's desperate housing shortage, they dare not withhold rent, thus risking eviction; neither dare they pay. The progress of the rent-man is watched carefully, and the rent books periodically inspected by the area vigilantes. Here is a new situation of intense conflict for the Catholic mother—a conflict between her desire to provide a stable and permanent home for her family, and her need to conform to group demands. Because, in an insecure, closely-knit group, the nonconformist gets short shrift indeed. It seems that she is threatened whatever she does. She knows, too, that the money will some day have to be paid, and that somewhere the debt is accumulating. But she has no cultural predisposition to long-term saving; immediate needs are paramount, and she knows that, however the debt is finally to be recovered, the money just will not be there. Social workers have been able to help some of these parents by themselves accepting the weekly payments and keeping the money until the campaign is over and the total has to be paid—but only a few can be helped in this way.

With the establishment of Long Kesh internment camp and as, towards the end of 1971, total arrests reached several hundred, there was added to earlier stresses the fear that the father could be interned. At the time of writing no charges have been preferred against internees, and the general impression in Catholic areas is that Republican sympathy, either now or in the past, is a sufficient criterion for arrest and detention. Consequently, few Catholic mothers have been able to watch the Army coming up their streets at night without at least a momentary pang of anxiety.

An Army swoop must, of course, be unexpected; it is often dramatic, and —at least for a child—always frightening. The mother's fear for her husband's safety is very quickly communicated. A ten-year-old patient said: 'They came up the street at five in the morning, breaking windows. They jumped on our stairs and broke them. I screamed and hid under the bed.

'A man I know is interned. I feel sorry for his girl and boy, who are at school with me. When I heard about it yesterday I couldn't stop crying.

'They might take my dad; they would come in the middle of the night. I often can't sleep for thinking about it and have to go into his room and see if he's still there. He hasn't done anything wrong, but neither had the others.

'The last time they came in they took some men away and as they were leaving they shouted, "We'll be back for the rest tomorrow." '

Soldierly banter perhaps, but the result was intense separation-fear in this child, as it is in many others. And for children themselves, there can be intimidation. Few sights can be more harrowing than the daily spectacle of files of youngsters flanked by soldiers going to and from school—Catholics on one side of the street, Protestants on the other.

The symptoms of these two children date from the outbreak of violence in 1969, but have been perpetuated through the bombing and intimidation campaign of 1970–73.

Paul is eight, and lives in Ballymurphy, an area constantly exposed to clashes between the security forces and hostile crowds. For nearly two years he has been having nightmares; he is still terrified of being in his bedroom alone, and his sleep and appetite have suffered badly. He often falls asleep in school, and has to be sent home by the teacher.

Even on the rare nights when Ballymurphy is undisturbed, any noise wakens him; on one or two occasions he has wakened at the sound of running feet in the street and screamed. He is afraid to sleep at night and continually becomes anxious about his own and his parents' safety.

His father has been receiving psychiatric treatment for symptoms of anxiety since the outbreak of the trouble. He 'talks compulsively' about it, according to his wife.

Paul himself is a small, underweight child. He is a non-identical twin; his brother is taller, stronger, and generally healthier. He is a boy of average intelligence, but his school attainments are very low—presumably because of virtual non-attendance over the past year.

He is still in treatment and is now finding it easier to express his fears through conversation, play and drawing.

Una, aged eight, was admitted to hospital towards the end of 1971, apparently unconscious. After being put to bed she woke up and began to shout incoherently about guns and bombs, and said that her father was going to be killed. She screamed, then lapsed into unconsciousness again. This pattern of waking, screaming, then becoming inaccessible continued for some hours. All physical investigations were negative; in due course she responded to sedation and was transferred to the child psychiatry inpatient unit.

She lives in a working-class Catholic district; her father has been active in an 'illegal organisation' for the past two years. Shortly after the outbreak of disturbances, the father of one of her school friends was shot dead by the Army. From that time she began to have screaming or crying fits if she heard gunfire or explosions or, later, any loud noise.

On the night of admission her father had gone out. At about midnight there were several bursts of gunfire unusually near, then the house was shaken by an explosion. Una screamed 'Daddy!' then her limbs went rigid and she fell.

Once in hospital her fits of acute anxiety subsided quickly. She was a shy little girl who did not find verbal expression easy, but she quickly became attached to a battered golliwog and, in therapy, was able to express her anxiety by 'telling Golly'. Soon she became able to talk more freely about her fears that her father would be shot or interned, or that their house would be destroyed. Her mother found it difficult at first to accept these expressions of fear. She herself was having intermittent fainting fits which seemed to be provoked by emotional stress. However, with encouragement, communication between mother and daughter became more ready and a degree of mutual support was attained. Una had two successful trial weekends at home; both she and Golly have now been discharged.

6

Disorders and Defences

'I heard guns. I was afraid. They were shooting guns at night and I couldn't sleep and I was afraid they would stick the guns in the window and shoot. So I slept with my mother. A store close to my window was burning. They were lighting matches and breaking windows and I thought the whole world was burning.'

Four-year-old girl in Watts, Los Angeles. 1965[1]

'The guns sometimes keep me awake all night. Sometimes I can't stop crying, because I'm afraid that the Protestants will burn all our houses or that my dad will be shot or interned.'

Eight-year-old boy in Belfast, 1971

There is a certain universality about a child's response to disaster. The varying realities of the event may well add details to nightmare and fantasy, but the child's fear is always, in essence, that of loss of the factors that make for physical and emotional security. He dreads the prospect of separation from his parents as much, if not more, than he does bodily harm to himself —an aspect of preventive psychiatry often forgotten in the rush to evacuate children from disaster areas. Later, the child's real needs usually become painfully evident.

But first—how do children *in* a riot or disaster area respond?

The symptoms of the ten Belfast children described in the previous chapter differed widely; problems brought first to the doctor's notice ranged from asthma to epilepsy. But, at the same time, all were broadly similar in that each closely resembled a classical *phobic anxiety state*.

'Phobia' is a fairly familiar word to most people. It means an intense, abnormal fear of some specific object; later, the fear may attach itself to other, similar objects and, in the end, the sufferer may be quite badly incapacitated—even to the extent of being unable to leave his house. A cat-phobic, for example, has a pathological fear of cats (pathological because it is out of proportion to any possible harm the cat might do to him), and, at a later stage, he may become fearful of other small animals, or furry objects, and finally may refuse to go outdoors at all for fear of meeting a cat. This process is known as *generalisation*. Symptoms of anxiety, at first related to a single object, later attach themselves to other objects which remind the sufferer, consciously or subconsciously, of the thing that was initially feared.

This is exactly what happened in the Belfast children who developed long-term symptoms even after the street disorders near their homes were over. Margaret, for example, fainted when she first heard gunfire. Later, she fainted when a car backfired or someone slammed a door. A rioting crowd and burning buildings precipitated Marie's epileptic fit; later fits were 'triggered' by a football crowd and a domestic fire. Soon she became afraid to go out because she was afraid of seeing crowds or fire-engines. The phobia had by then generalised.

But to describe and analyse symptoms in a small group of children is scarcely an achievement in itself. All these young patients lived in densely-populated areas; thousands of other children experienced exactly the same sights and sounds—the smoke, the gunfire, the angry crowds. Why, then, did only a small minority develop crippling symptoms of anxiety? This is really the crux of the problem. Identification of the factors that lead to some children breaking down, but not all, may make us more able to find ways in which children may, in the future, be protected from this type of reaction. There may also be answers to questions about helpful intervention and treatment.

Thus behavioural definition of these children's symptoms, such as a 'conditioned fear response', is only a starting-point. It would be just as sterile to put the symptoms down to deeper intra-psychic conflicts. Writers who have done major work on phobias believe that their cause is usually multiple.[2-4] Lindy Burton, dealing with nervous and behavioural disorders in children following traumatic events, reaches the same conclusion.[5]

In Belfast, too, the way in which each child reacted to riot stress seemed to depend on three main factors. There was first, the degree of emotional security enjoyed by the child both before and during the period of acute stress. This related not only to his *own* psychological resources, but also to those of his immediate family. Secondly, there was the role of the stressful experience *itself*. Thirdly, each child's response was idiosyncratic, or unique, depending on his own usual way of responding to new experiences.

Taking the first point, the most outstanding common factor was that of *vulnerability*, not only in the patient himself, but also in his parents. In each child there were all of the following: an earlier tendency to nervous symptoms, lack of physical robustness, and a tendency on the part of one or both parents to over-react to the threatening situation. In all cases, in fact, at least one parent was suffering from incapacitating symptoms. Typically, there was a mother who would become acutely agitated during a riot in the street, or when the child mentioned his fears—and a father who would respond, in the same circumstances, with anger and aggressive behaviour. The effect was that the child, perceiving his parents as vulnerable, felt inhibited from expressing his anxiety—anxiety which later communicated itself as a psychiatric symptom. No child was 'disturbed' in isolation; each problem, on examination, proved to be that of a disturbed family. The parents' inadequacy when it came to providing emotional support had been evident long before the period of acute stress; then it had broken down completely. But this role was a crucial one. One child (not in the series) said: 'My parents were calm, so I didn't worry. If *they* had been frightened, it would have been awful.'

Thrown back thus on his own resources, the child's response was *adaptive* in that it was his major means of conveying distress; it also served as a defence in that it *altered the degree of contact with the environment*. This is clearest in the children who had fainting fits—'switching off' a reality which had become intolerable. The symptoms were also adaptive in that they continued, in each case, until attention was drawn to the child's plight and he was able to escape from the environment which he found threatening. When this had been achieved, or when he had an opportunity to express and discuss fears freely, the symptoms diminished or disappeared.

Why did the symptoms differ from one child to the next? This depended heavily on each child's own personality—his strengths and weaknesses. So Anne, the histrionic girl, developed dramatic hallucinations, Margaret and Maeve, fearful children, had fainting fits, and Peter, with his long tendency to somatic, or physical, symptoms of anxiety, came to us with incapacitating nervous system and gastric complaints. It was as if each child had his Achilles' heel—the point at which distress showed when the immediate family failed to respond adequately to his needs during a period of acute stress.

It is interesting that these were all children from approximately age eight to puberty. This clustering of ages is probably related to levels of comprehension. Younger children did not fully understand the danger; the truth was less evident to them. After a nearby explosion, a mother could often reassure a child by saying, 'It's just thunder.' One mother told me, 'He used to be terrified of thunder, but now he just goes to sleep again.'

(A researcher in California, after the 1971 earthquake, found that for young children an 'explanation' such as 'It's just a truck passing' had in many cases been adequate reassurance.)

On the other hand, older children were more likely to find refuge in action or flight, and were less susceptible to wild rumour—such as the tales that evoked Sean's symptoms or Ann's.

So far, these major questions arise. Can the three features under the heading of vulnerability (earlier physical and emotional ill-health and parental inadequacy) be taken as the main predictors of psychiatric illness in children exposed to major stress or disaster? If so, is the converse true —that a previously healthy, well-adjusted child in a stable family is unlikely to develop persistent symptoms in this setting? And, most important, can children at risk be identified at an early stage and their symptoms prevented or alleviated?

Results from Belfast suggest affirmative answers to all these questions. Is the suggestion—and it can be little more at this stage—supported by the results of work with children in other, similar, situations?

Some war-time studies are of interest, although they deal only with acute symptoms; no children were followed up for more than a month or so. Also, as has been suggested in Chapter 4, wars and riots differ in major respects —not the least being that of their impact on children. (Having myself weathered the London blitz as a very young child, I can make this point with an extra degree of confidence.) War-time children rarely saw the enemy; they didn't see the injuries, the blood, the dead, the sheer naked aggression of battle. And for them, of course, there was no active participation in attack or defence—as there is for Ulster children. 'Battle stress', though a well-recognised symptom in military psychiatry, is certainly a rare phenomenon in children.

During the Second World War, in fact, emotional disturbance in children was not particularly prominent. For example, following a major air-raid in Bristol, some 60% of children were found to have nervous symptoms, but these persisted in only about 10%, all under the age of five. The reaction of all children to subsequent raids was comparatively slight.[6] In the same year, a report on children evacuated from Manchester during air-raids indicated that there was much greater evidence of anxiety in reception centres than in heavily-bombed areas. It was suggested that the emotional strain of removal and separation was greater than that of the threat of bombing.[7] A London psychiatrist identified the condition of 'air-raid shock'—the presence of acute nervous and behavioural symptoms in children following bombing and evacuation. The longer-term outlook for the children affected was generally good.[8]

In 1952, the World Health Organisation published a report by Bowlby, reviewing the effects of war experience and various other kinds of childhood disasters. He emphasised the harmful effect of separation from parents—a common aftermath of war.[9] Several other investigators commented on how, during the blitz, children did not develop symptoms of anxiety until they were evacuated to areas not only peaceful, but materially far superior to their own homes. Bed-wetting, sleep disturbance and behaviour disorders were common, and these symptoms often did not remit until the child was re-united with his family in the war-battered City.[10]

The Cold War of the fifties and sixties seems as remote now as the World Wars did then. But at the time, several investigators found a surprisingly high level of awareness among children of the threat of nuclear war.[11] [17]

During these two decades, older children and more intelligent children were generally better informed, more optimistic and less susceptible to rumour. In 1963, for example, Adams found that the amount of anxiety expressed about the possibility of nuclear war diminished steeply from the ages of 10 to 19. Escalona, also in 1963, reported that, while younger, grade-school children mentioned many worries and often asked questions of their parents about the war threat, the older children were more factual and realistic about the possibilities. Allerhand, in 1965, also found that younger children were constantly seeking answers, and added that the presence of 'comfortable and receptive adults' was critical to their emotional stability.

This was, in fact, the major point of agreement to emerge from all the Cold War studies—the central role of parents and relatives in both providing information and coping with anxiety. Darr (1963) concluded that the child's reaction depended on the adult environment through which the threat was filtered down to him, and that his reaction closely mirrored his parents'. A parent could, on the one hand, conceal his anxiety unduly to the point where, while the child felt secure in his home environment, he could often, when outside it, meet anxieties with which he was ill-prepared to cope. On the other hand, parents who expressed helplessness seemed to provoke the greatest degree of child anxiety. The most constructive approach seemed to consist in a sharing of anxiety between adult and child in the common belief that something could be done to meet the threat. The child was thus helped to deal with his anxiety *within* the reality. Several others of the writers I have quoted also used the 'mirror' metaphor in discussing child reactions as related to parents'.

In 1965 Elder, having dismissed the mass of anecdotal material that bedevils such subjects (it happens elsewhere than in Belfast), reviewed the results of all systematic observations. He concluded that, of all potential determinants of childhood reaction to the threat of war, those in the category of parent–child relations were the most important. Wrightsman (1963) found a definite positive relationship between the extent of parents' discussion of war and the child's worry about war. But the important variable was not just the *amount* of talk about war, but the degree of anxiety that accompanied it. At the time of the Cuban crises, Schwabel found the same trends in a sample of three hundred schoolchildren.

Elder's final interesting observation was that, unless it was conveyed to him by adults, a child might not otherwise experience anxiety in a threatening situation. His parents have known a different situation, and to them the present one seems bizarre; for a child it might be commonplace. This might partly, though I think not entirely, explain why in Belfast, with an experience of some five years of street violence, anxiety is much less prominent in children under the age of eight.

(At the time of the Kennedy assassination, awareness and feelings of involvement among American schoolchildren were unexpectedly prominent; parents tended to underestimate their children's concern. Each child in some way felt implicated, and took up his own defensive position.[18-20])

A child's reaction to natural disaster seems to have the same predictors as his reaction to war and rumours of war. A natural disaster is, by its nature, unexpected; detailed or sophisticated studies are rarely possible. Further, since most of this work has been done in the last few years, there is little in the way of follow-up information available. The studies at present available are most useful where they illustrate how symptoms vary with age and family structure.

After a tornado in 1953, Perry *et al*[21] found anxiety reactions and regression to more infantile behaviour in most school-age children; Crawshaw[22] differentiated clearly between the reactions of children of different ages. Those under eight directly reflected the parental response to the cyclone. Children of 10 to 13 often displayed excitement with no conscious experience of fear, while teenagers expressed more conscious anxiety with the excitement. He believed that this excitement represented a wish not to take the threatening reality seriously. After the Crimean earthquake of 1927, Brussilowski also found conscious anxiety most prominent in the 13 to 17 age-group.[23]

In October 1966, Aberfan, South Wales, was the scene of one of the most terrible disasters of this century. A collapsing coal-tip engulfed a school, killing 116 children and 28 adults. Dr Gaynor Lacey, a child psychiatrist, treated 56 children who developed emotional symptoms following the landslide. Nineteen of these had lost brothers or sisters in the disaster. The most prominent symptoms were sleeping difficulties, nervousness, bed-wetting and emotional instability. All of these children had *other* anxiety-creating factors in their environment (such as an absent or inadequate parent). Long-term symptoms were uncommon, but most of the children had abnormal fears of weather for a long time after the disaster. (Recent reports describe an uncannily similar pattern of childhood symptoms after a flood in Buffalo Creek, Virginia, when, in February 1972, a coal-slap dam gave way and inundated a small mining community, killing 125 and leaving 4,000 homeless.)

The story of one small Aberfan boy, when I heard it, reminded me vividly of my own disturbed Belfast group. He developed encopresis (faecal incontinence) on the evening after the landslide. This soon cleared up, to recur months later, one wintry night, when he was wakened by the sound of snow sliding from the roof of his home. In the same way, symptoms with their onset at the time of the acute stress were later re-evoked in the Belfast children by stimuli which resembled those experienced on the first occasion.

In 1971, Dr Howard Hansen reported in detail on the effects of the California earthquake on 41 children, ranging in age from 2 to 17 years.[24] Their predominant fear was, he found, of separation from parents and loved ones. Another group, working in the same area, also reported separation from parents as the most prominent fear in children of all ages.[25] Hansen's other main observation was that the reactions of children who were with their parents at the time of the event usually mirrored those of the parent. He comments: 'Since a child bases his view of the external world largely on the thinking of significant adults, it is not surprising that in time of danger he looks to them for a clue to interpret the situation. Such is the parental influence that however imminent or serious the danger, the child may be only minimally disturbed if his parent remains calm and dependably supportive. Conversely, the fear and anxiety of a catastrophic experience may be heightened for him by his parent's behaviour.'

Even when the parent was not actually with the child during the earthquake itself, it still seemed that the pre-existing pattern of relationship between parent and child invariably influenced the child's response to the stress situation.

As regards age differences, pre-school children, as in other studies, had little awareness of the realities and thus little anxiety. Latency children (7 to 12), understanding the physical dangers more clearly, showed more anxiety and greater need for reassurance. Pubescent children, still more aware of possibilities, found the event most threatening of all groups.

Hansen believes that the most crucial determinant of response to disaster occurs during the period immediately *following* the disaster, when the need for ventilation, usually by talking, is strongest. But, for many adults, recollection of the earthquake raised such anxiety that they tried to suppress children's attempts at discussion or questioning.

As time passed, other modes of expressing anxiety became apparent. (In Aberfan, after the landslide, children played 'burying' games.) Telling stories about the disaster and drawing pictures also were common means of expressing anxiety.

A surgeon told me of an occasion in the Second World War when about a hundred members of the WRNS, rescued from a sinking destroyer and mostly with acute emotional reactions, were asked to write an essay on their experience. This had in all of them a strong anxiety-reducing effect (especially in one who sold her story, for a substantial sum, to a Sunday newspaper).

Longer-term features in the California children exposed to the earthquake were recurrent 'catastrophic drawing' and exaggerated sensitivity to external stimuli reminiscent of the original traumatic experience. In other words, Hansen found much the same reaction as we have seen in Belfast children who faint when a door slams, or in Aberfan children who become acutely anxious in heavy rain.

Hansen concluded that the children's main emotional needs were human company, physical closeness, information and, later, a chance to question and discuss the experience. These needs, he believes, can be met by nonprofessional volunteers.

Howard,[26] working in the same area, reported similar findings. The children who did best were those whose parents *accepted* the regressive behaviour and the child's need to 'talk out' the experiences. In groups, children needed to find their own counter-phobic defences (e.g. out-doing other children's experiences by boasting about their own). Howard, too, recorded some longer-term symptoms; a few children insisted on sleeping on the floor for several weeks after the earthquake.

There appears to be only one detailed study of children's responses to a series of riots, other than in Belfast. In 1965 in Watts, Los Angeles, three days of rioting in this predominantly Negro district resulted in several deaths, a few hundred injuries (exact numbers vary between different accounts), and damage to property extending over fifty square miles of the city.

Structured interviews with 197 Negro, 23 Mexican-American and 52 White children revealed a high awareness of events in all groups, although, naturally enough, reaction was attenuated by distance.[27] Emotional defences were prominent: there was a high degree of *denial* of fear: that is, only about a quarter of the children admitted to having been 'scared', but this proportion rose to over one-half when ratings were based on *objective* signs of anxiety reported by parents and teachers—for example, tears, tension, praying, sleeplessness.

The observers found high awareness of racial difference in all these children, and observed that there exists, in particular in White children, an 'almost uniformly hostile-fearful attitude towards Negroes, amounting to a subcultural stereotype that Negroes are by nature bad and violent. There are important, and unhappy, implications when a stereotype like this appears so clearly by the age of four or five. If this small sample is at all representative for young Mexican-American children, the possibilities for future cooperation and understanding between the two minorities are indeed dim. The finding suggests that if inter-group attitudes are to be improved, efforts to change them must begin at a very early age.'

The group comment finally that, while enduring effects of the riot experience are likely, no objective proof of this is available.

Indeed, there is virtually no literature on the long-term effects of violence on children. There is some evidence, however, that some distortion of perception becomes apparent, particularly of concepts of war and peace. Non-conflict, as a goal in itself, may be greatly over-valued. An American physician dealing with a civilian population in Vietnam wrote: 'The war has been going on for so long that no one under 35 has any real memories of what things were like before. The children think that "Peace" is a beautiful time and place where all problems are solved and everything is lovely —something akin to heaven on earth.'[28]

By contrast, a study of six thousand children from peace-time countries found concepts of peace generally weak. To these children peace was mainly negative, the absence of war.[29]

Children respond in a strikingly similar way across the whole range of wars, riots and disasters: the crisis colours the reaction without defining it. The common pressures—poverty, lack of family cohesion, overcrowding and malnutrition—provoke common syndromes, almost regardless of the event that finally pressurises the family beyond its ability to cope.

On the basis of available results, we can now try to construct a theoretical and practical framework for dealing with children and families under acute stress.

Stress is, of course, a tricky word to define; it can sometimes mean the event itself, and sometimes the reaction to it. And what is stressful to one person might be an everyday occurrence to another—taking a plane-flight, or making a speech, for example. This could be criticised as an academic quibble in the context of riots or disaster, which are likely to be anxiety-provoking to almost everyone: still, stress must be defined as an entity before its effects are evaluated. So, as a working definition for the present purpose, a 'stressful event' is taken to mean *one which makes unusual demands on physical or emotional resources.* The value of this definition is that it takes account of individual differences.

Are similar means of treatment and prevention as effective in Belfast as in other disaster areas? Can we learn anything from surveys and treatment methods used elsewhere? Or (with due modesty) can workers in other areas learn anything from us?

To begin with, consider this sequence:

'I thought I was going to die.' 'I was terribly scared . . . I thought the whole building would go down.' 'I had to have someone to talk to . . . the nurse went out . . . I remember praying, "Please God . . ."' (Children's memories of the California earthquake, 1971[30])

'He can't think about anything but the troubles. He's always asking questions. If there's any noise at night he wakes and starts shaking, and has to know what it is. Talking about it and going on about it all the time, he is.' (Mother of ten-year-old Belfast boy, 1971)

'As a child of five–eight years at the time of the troubles in the early 1920s, I can remember my personal reactions only too well . . . If my parents were late home, even for a short time, there was always the deep fear that they were dead . . . I still jump at a loud noise, even a car backfiring, after all these years . . .' (Reader's letter to an English newspaper[31])

These quotations may help to illustrate what is, I think, a useful way of thinking about the effects of stress. In Belfast, as in other disaster areas, a child's response can be divided into three distinct phases; the child has different needs during each phase.[32–40]

During the initial phase—the *period of impact*—most investigators agree that, while about a quarter of the people remain calm and capable of purposeful action, another quarter suffer symptoms of acute anxiety such as weeping, shaking fits, and the remainder, the majority, are partly immobilised by inertia. These latter are dazed and stunned by an event which they cannot yet comprehend. All of these reactions were reported in Belfast immediately after riots and explosions (Chapter 4), although it would be impossible to state proportions with any precision.

Among children, the state of general, purposeless excitement evident during the California earthquake had its counterpart in Belfast. Teachers, play-group organisers and a group of American student volunteers told me how, during or after a period of rioting in their area, the children seemed to be possessed of intense and ill-directed nervous energy, shouting, singing, dancing about, giggling inanely.

Several observers have interpreted this behaviour as evidence that the youngsters were 'enjoying' the new situation. But careful assessment of children with this type of reaction, both in Belfast and in Watts,[27] showed that these children were, in reality, concealing a good deal of anxiety about their own and their parents' safety. So it is important to recognise this 'denial' reaction for what it is. Apparent excitement and laughter in a stressful situation do not mean that a child needs no reassurance—rather the reverse.

Reassurance means, initially, giving factual information. Most of the writers quoted have emphasised the damaging effect that wild rumours can have on a child's perception of a situation. In my own series, for example, Sean and Anne reacted violently to the grossly inaccurate accounts they heard on the streets as there was no factual information to counter them. During the California earthquake, hospitalised children followed the nurse around begging for information, and seemed disproportionately grateful for any that was given. Some had believed that the world was sliding into the sea. Children of four and five in Watts thought 'the whole world was burning'. Accurate information invariably reduced anxiety. Why? Shielding a child from the facts is usually self-defeating, since his fantasies tend to depict the event as much more terrible than it really is. Early acquaintance with the truth is the rational way to prevent this happening.

Besides reassurance, a child has an intense need for physical closeness to trusted adults, usually his family, during the period of impact. This explains why a child usually fares better emotionally in the riot situation *with* his parents than out of it and separated.[9,10] Unless there is immediate physical danger, all available evidence seems to show clearly that early evacuation of children is not indicated. If it is, families should be evacuated as groups. Unfortunately, many of the mass evacuations from both Catholic and Protestant areas of Belfast[41,42] have been nothing more than irresponsible pieces of political window-dressing, and the children have suffered needlessly.

In August 1971, for example, about two thousand people, mainly children, streamed south to Dundalk and Gormanstown. A psychiatrist[43] who spent a great deal of time in the resettlement areas told me about the children's clinging, regressive behaviour. Their need for even a limited degree of group identity was intense, he said, and friendships were jealously guarded. Once, when it was proposed to move a child's bed, the others stood round it and refused to allow the child to leave. On an outing, groups of children would not come out of the vans in which they had travelled until assured that these *ad hoc* groupings would not be broken up. Having sought and found a surrogate family, the youngster was determined to hold on to it by all the limited means at his disposal.

When the immediate crisis is over, the *period of recoil* is marked by a need to talk about the experience, sometimes in a repetitive, compulsive way, a process often called ventilation or talking-out of anxiety. It is as if the child is re-living the event in fantasy, in an attempt to experience once again, and finally master, the anxieties he could not cope with at the time.

Of course, for a young child talking is not the only, or even the most natural, means of communication. Re-creation of experience in play and drawing is universal child behaviour. Taking thus a more active role, he can then to some extent control the experience, and so feel less anxiety. Children played 'disaster games' after earthquakes and floods, and 'burying' games in Aberfan. In schools and playgroups after riots in Belfast, Watts and Harlem, children have built barricades, divided into groups and thrown missiles.

But in older children (by which I mean approximately age eight to puberty), the need for verbalisation coupled with adult reassurance is vital; mishandling at this stage can lead to nervous symptoms later. It seems that a fear that cannot be communicated in any other way is likely to be expressed in due course in the form of a nervous symptom—bed-wetting, nightmares, fainting fits.

The results of all studies of children in this phase can be crystallised in a few words: the child takes his reaction-cue from the adult figure nearest to him. If the adult displays undue anxiety and thus cannot tolerate, in addition, the child's expression of his own fears, this valuable avenue of communication is blocked, and the child's risk of suffering long-term disturbance increases. Role-reversal, for example, reported in several disaster situations, was also a prominent feature in our Belfast group. Here, the parent uses the *child* as a psychological prop, pouring his own anxieties into the child's ears.

A child, sometimes, may have to be actively encouraged to talk about his fears, so that rumours can be discounted, and misconceptions corrected. An indirect approach is sometimes useful here. For example, wild horses often would not induce a small boy to admit that he was frightened—yet he continues to have nightmares and eneuresis. Useful leading questions are: 'Was anyone *else* afraid? Was your sister afraid? What was she afraid would happen?' Giving answers to these questions allows the youngster, without loss of status, to describe his *own* fears and to receive reassurance.

The *post-traumatic phase* of response to stress has already been discussed; this is the period when there are fears about a recurrence of the disaster, and high awareness of environmental reminders. All the children described in Chapter 5 were in this phase; it seems that it is particularly prolonged and incapacitating if there has been mishandling in the earlier phases—that is, if there has been lack of support and of opportunity for discussion. The most effective treatment of anxiety symptoms at this stage consists of a *recapitulation* of the phase of recoil; the child, in therapy, is given the opportunity to express his fears in talk, play and drawing. Gradually, he comes to relive and integrate the stressful experience, this time in a calm and supportive setting; with learning of new responses, psychological growth takes place. Both parent and child are involved in therapy, the aim being to facilitate communication from within the family, so that no member needs recourse to psychiatric symptoms to express anxiety or to request help.

After the Los Angeles earthquake, one psychiatrist[26] found that, in group therapy, many children only required a single visit for relief of symptoms. In Belfast we cannot be optimistic enough to hope for such dramatic results. With the long chain of referral from family doctor to paediatrician to psychiatrist, disturbed Belfast children frequently are not seen by us until symptoms have been present for more than a year, when they are likely to prove much more resistant to treatment.

What can be done? The need for emergency provision of material resources in a disaster area is well recognised, but provision of psychiatric services lags far behind. At the same time, the vast majority of emotionally disturbed children in this type of situation do not need the services of a psychiatrist as such. The therapist, indeed, is usually the parent, unrewarded but supremely effective. Very many other children can be adequately supported by GPs, Health Visitors, relatives and voluntary workers.

With professional assistance severely limited, the major task of an emergency team is to identify families at risk so as to direct preventive measures where they are needed. From then on, most studies have stressed the important role of *non*-professional helpers.

It is largely to this group—to parents, teachers and other adults faced with, and perhaps nonplussed by, the task of coping with adults and children in a disaster area—that I want to direct the following brief resumé.

Acute anxiety is common during or after a stressful event, such as a riot or natural disaster. It can take bizarre and sometimes frightening forms —shaking, trembling fits, laughing for no apparent reason, uncontrollable weeping. This is not the same as a nervous breakdown or a mental illness; it is a normal reaction and rarely lasts for long. In children, it often takes the form of bed-wetting, sleep disturbance, and repeated requests for reassurance. In the acute stage, physical closeness to a trusted adult is essential —an adult whom the child can trust to remain with him as long as he requires it, who can accept without comment his reversion to childish, sometimes aggressive behaviour, and answer patiently, again and again, the same questions. It must be remembered, too, that apparent excitement, giggling and laughter for no obvious reason, can conceal deep anxiety.

A child's *major* fear at this stage is of separation. As he can suffer as much from separation from his family as from the stressful event itself, evacuation should not be considered unless there is physical danger, in which case each family should be evacuated as a group.

Adults and children need factual information; they are often treated impersonally in a disaster situation, and no facts are given. Information is especially important for young children, whose imaginations usually depict a disaster far greater than the reality.

If acute anxiety symptoms persist, a doctor is most likely able to help, perhaps by prescribing a tranquilliser.

At a later stage, children who continue to show anxiety symptoms need ample opportunity to express specific fears and, perhaps, to re-enact the experience in talk, drawing and play so as to enable discussion and support. Parents need reassurance that any unusual degree of dependency is transient, and that this phase is normal and therapeutic rather than harmful. The role of confidant can often be filled amply by a relative, friend or teacher.

Children between the ages of eight and twelve, children with a previous history of physical or emotional illness, and those from unstable homes, are particularly at risk.

We have so far concentrated on similarities—and quite properly. In the context of emotional illness comparisons across different cultures and disaster areas are clearly valid. But there are fewer precedents to help in dealing with Ulster's additional, much larger, group of problem children—the aggressors. Slowly, painfully perhaps, we have to find our own answers.

The next two chapters attempt to unravel the chain of cause and effect that finally presents us with that most dangerous of fauna, the child guerrilla.

7

Republicans and Sinners

The Ulster conflict is not a holy war. Most recent writing about the Province implies, essentially, a domestic squabble between Protestants and Catholics. But a sharp religious, physical or even dietary difference may distinguish two parties to a dispute without that difference being necessarily the major irritant. In Ulster, while it is true that most of the people on one side of the Peace Line are Protestants and most of those on the other Catholics, it is still a wild leap of logic to Press leaders that read 'Shooting and burning in the name of Jesus . . .', 'Killing for Christ . . .'

Besides being misleading, this style of reasoning can also be incongruous. Thus the introduction to G.W. Target's *Unholy Smoke*: '. . . a fist in the teeth for your religion. Throw a stone for Jesus, hurl a petrol-bomb for God, shoot a Protestant for Holy Mother Church . . . A child dies, shot by a bullet meant for a man—a gift from one Christian to another.'[1]

Target loses a degree of credibility, anyway, by making his Shankill Protestants talk like characters out of *Riders to the Sea*: ' . . . but knowing our Fenian gentlemen only too well, we were knowing there was only one reason why they'd be wanting to be on the roof of that school—to be pouring a hail of bullets into this street . . . Catholics don't get the education, it's labourers they are. What else can they be doing without the education . . . So you don't be hoping anything extra from their sort than what they did, roughs from the digging jobs, digging drains. They were like mad things that night, and fighting drunk the most of them . . .'

No Ulster Protestant, he of the tight-lipped Puritan ethic, talks in this discursive, Western-Irish style. And here, precisely, is the point being missed: Ulster's problem is a *racial* one, a conflict of cultures and ideals between two ethnic groups as distinct as are Blacks and Whites in the United States or Southern Africa. In at least one respect the two groups are, in fact, even more distinct, since mixed marriage, always the exception in Ulster, is now so uncommon as to rate Romeo and Juliet treatment in local news media.

Community inbreeding, just like inbreeding in a family, emphasises and exaggerates unique characteristics. The end result in Ireland, after three centuries, is two races that contrast physically, emotionally and ideologically. At extremes the Celtic group[2] have darker hair and eyes, more swarthy complexions and more angular features than their Anglo-Saxon counterparts, as well as an accent with slightly longer vowel-sounds and softer consonants. This accent difference, as well as being an expression of varying geographical roots, may also reflect something of the importance laid on the Irish language in Catholic schools.

These physical and accent differences, although by no means universal, are still sufficient to enable most Ulster Catholics or Protestants to identify one another on sight. If not, other clues remain. Catholics' surnames, of native-Irish origin, have O' and Mac as usual prefixes, and Christian names are almost always those of Irish saints or heroes. Protestants' names tend to be solid Anglo-Saxon or Scottish. Willy John, a traditional Protestant figure of fun, has his opposite number in the Irish-Catholic P.J. (Patrick Joseph). If an Ulsterman's religion still cannot be established, his address will leave little doubt, and the name of his school none. Schooling is totally segregated, and housing soon will be.

At school Catholic and Protestant children play different games—Gaelic and English games respectively (which makes even the possibility of integration on the sports field remote). Catholics have larger families. The Church's ban on contraception and, indeed, its positive approval of multiparity, makes a large family culturally almost inescapable.

So these differences add up to a high *visibility* between the two ethnic groups. In a 1970 survey a major employer was asked how many Catholics he had on his staff. He replied that he didn't know. 'We don't ask prospective employees their religion.' Although calculated to imply a liberal policy, this reply would attract as much Ulster scepticism as if the question had been about the employment of Negroes. (It turned out, by the way, that the firm employed no Catholics.) However, the Northern Ireland Commissioner for Complaints found this a stock answer to the question, and added: 'It is . . . unconvincing in the Northern Ireland situation, for the head of a small office . . . to claim not to know pretty accurately how many of his staff are Protestants, and how many are Roman Catholics . . .'[3]

I have chosen the analogy of the Negro in employment carefully. In this chapter, proceeding from the idea of two disparate cultures, I hope to develop an hypothesis that the position of the Catholic in Ulster is virtually identical with that of the Negro in the USA or Southern Africa and, to a lesser extent, that of other minority ethnic groups. He suffers from the same social pressures and stereotypes—and responds with the same forms of protest. Further, I shall suggest that it is only in this context that the Northern Ireland conflict can be fully understood, and in which the most realistic solutions can be propounded.

The Irish Celt's subordinate position in Anglo-Saxon culture is not, of course, new (although the concept, to my knowledge, has not yet been specifically applied in Ulster). As long ago as 1880 a Belgian essayist, Gustave de Molinari, wrote that England's largest newspapers 'allow no occasion to escape them of treating the Irish as an inferior race—as a kind of white negroes . . .'⁴ In *Harper's Weekly* in 1876, a cartoonist depicts an Irish immigrant and an emancipated Negro slave equally balancing North and South votes during reconstruction; *Puck*, an English satirical magazine, printed a cartoon depending on a play of words between 'Ashanti' and 'a shanty'—implying a parallel between Pat in his Irish hovel and the African in his mud hut, reinforcing, in a reviewer's words, 'the ties between Irish Celts and Black Africans'.⁵

The same blend of racism with the devastating crudity of mid-Victorian satire is the stock-in-trade of Belfast's *Protestant Telegraph* (29 May 1971): 'It is rumoured that . . . a young soldier was injured very painfully by a missile thrown by one of the "persecuted minority". As he lay on the ground, groaning and writhing in agony, the "Irish negroes" danced around him, singing and laughing.'

Or (2 October 1971), 'On Saturday, a mob of Apaches from the Markets area headed towards Short Strand, where they picked up another tribe . . .'

And, in the *Loyalist News* (28 August 1971), '. . . these white negroes . . . sub-creatures . . .'

In a letter to the *Belfast Telegraph* (14 August 1971), a Rhodesian offered scathing comment on 'the white trash of the Falls and Ballymurphy'.

Little enough, so far, to provide a starting-point for the argument. But Irishmen and Negroes have not, of course, been the only great outcaste groups in history. Jews have suffered the same fate for even longer; indeed, the Jew's position, especially under the Roman Empire, has often been compared with that of the Irishman under British rule. Both, inhabitants of a relatively small nation, were invaded and scattered by an expanding, colonising Power. Both, maintaining a strong sense of nationality whether at home or abroad, have clung tenuously to traditional religious faiths; at home, despised by their conquerors as sensitive to insult, quarrelsome, music- and literature-loving rather than work-oriented; abroad, by contrast, often accused of too vigorously making their way in commerce, soldiering or politics.

One social scientist, taking this a stage further, drew attention to the differing psychological make-up of the Teutonic and Jewish races.[6] Their characteristics clustered, in each group, into *der Typus* and *der Gegentypus* respectively. The former had a relatively simple culture, was intolerant of ambiguity and mentally rigid. Der Gegentypus, on the other hand, had poor ability to think logically and was morally flexible.[7] These images, valid or not, have often been applied, though less formally, to the Anglo-Saxon and Irish races respectively.

How does this in-group/out-group concept apply in Northern Ireland? How is its philosophy and practice passed on to children, and how do they respond?

Every culture to a greater or lesser extent needs an outcaste group. Human society is based on caste, with status defined as much by group membership as by individual ability. A group, to maintain status, needs to perceive some other group as inferior; the resulting interplay of pogrom and protest is universal.[8]

But some societies have had a stronger need than most to define a 'pariah' group. Nazi Germany and modern White Africa are the best-known examples of this; now we can add Ulster. What special factors, other than the mere fact of a dual or plural society, operate?

It seems that the out-group emerges most clearly in the setting of an expanding economy when, with the growth of industry and opportunity, the pressure of upward social mobility disturbs a middle-class élite. This growing middle-class, as status for the worker rises, needs to find a new group to occupy the bottom of the hierarchy. The lower socio-economic groups too, their present poverty more evident by contrast with the new goals that a growing economy has made visible, polarise as competition to reach the goals increases. And they need a scapegoat to blame if those goals are not reached.

Thus it is that attempts to define and depict a common ogre or scapegoat meet with acceptance at this stage in a society's development. Thus it was, in Victorian England, that the Irishman became the subject of vicious attack, not with bomb and gun but, in the English style, with satire and cartoon.

'A gulf, certainly, does appear to yawn between the Gorilla and the

Negro. The woods and wilds of Africa do not exhibit an example of any intermediate animal. But . . . philosophers go vainly searching abroad for what they would readily find if they sought for it . . . in some of the lowest districts of London and Liverpool. It comes from Ireland . . . it belongs in fact to a tribe of savages; the lowest species of the Irish Yahoo. When conversing with its kind it talks a sort of gibberish . . . The Irish Yahoo generally confines itself within the limits of its own colony, except when it goes out of them to get its living. Sometimes, however, it sallies forth in states of excitement, and attacks civilised human beings that have provoked its fury . . .' (*Punch*, 18 October 1862)

It was in Hitler's expanding post-war Germany that the Jew became the focus of intensive harassment and economic pogrom; anti-Negro feeling in the Southern States grew with the transition from an agrarian to an industrial economy; and in South Africa, as mixed communities grow and prosper, the Black finds himself robbed of his place in the new meritocracy by the simplest of all processes—apartheid. The ancient Jews, themselves, outlawed the Samaritans as Israel groped its way back to economic viability after a ruinous series of wars. (If Jesus were reincarnated on the Shankill, who can doubt that he would tell the Parable of the Good Catholic? And who could not predict the reaction?)

Sadly, the priest, the Levite and the innkeeper are the more convincing archetypes. The upwardly-mobile citizen reluctant to share his new status with a largely agrarian, indigenous group is a familiar figure. A member of the old élite could afford to be generous to local workers; as the source of economic support and technical know-how his position at the top of the hierarchy was secure. But the threat of displacement presents itself much more clearly to second, third and subsequent generations—particularly those in the lower middle-class, who have to compete most fiercely for places in the new industries. Members of militant discriminating movements, such as the Ku-Klux-Klan, the White Caps and the Minutemen in the USA, the Racial Preservation Society in England, and the UVF and Vanguard in Ulster, tend to come from this group, rather than from working-class stock.

Under the title 'The White-Collar Gunmen', a London *Sunday Times* article ran: 'Exactly a week ago, an Ulster businessman whom we shall call Grant drove . . . to a quiet spot . . . outside Belfast. There he met a friend of a friend, who handed him a small box. It contained a . . . Browning automatic pistol and 50 rounds of ammunition.

'Grant is far from a hooligan; his greying hair is neatly brushed; his clothes run to well-cut suits and sober ties . . . Grant and six friends . . . have formed a group . . . Already they have begun to earmark

Republicans in their locality for selective assassination . . .'[9]

Army personnel believe that this report may be something of an exaggeration. At the same time, the bulk of the audience at militant Ulster Vanguard rallies have been, noticeably, well-dressed members of the lower-middle class.

Up to the present, there has only been sporadic recourse to physical violence against Republicans in Ulster. This is because, by virtue of its position, any dominant group has more subtle, and much more effective, means to establish and hold relative positions.

'This we will maintain'—Orange Motto.

To maintain the *status quo*, the out-group has first to be clearly defined; this entails the formation of a *stereotype*. Originally referring to a printer's mould for molten lead, this word now means, broadly, the conception of all members of a group as possessing certain stock characteristics. The idea is familiar, and runs through all sections of society. Professor H.J. Eysenck writes: 'Perhaps the most obvious field in which stereotyped attitudes are found is that of national differences. It is not, however, the only one. We all have mental images of certain groups of people which makes us endow these groups with certain uniform characteristics. Sometimes these characteristics are picked out in cartoons; the *Daily Worker* capitalist with his top hat, morning coat, and a bag of gold, grinding the faces of the poor in the dust, finds its counterpart in the *Daily Express* cartoon of the unshaven Bolshevist, bomb in hand, who threatens to blow up Parliament. Old maids, mothers-in-law . . . Jews, Nazis, workers, yokels . . . there is scarcely any large group in society which does not assume some stereotyped characteristics which are attributed to all its members, however unrealistic and inappropriate such attribution may be.

'But it is in the field of national differences that stereotypes appear with particular virulence . . .'[10]

True—and the interesting aspect of this is that the Negro in the USA or White Africa and the Ulster Catholic seem to share virtually the *same* stereotype—the in-group brand him as dirty, feckless, happy-go-lucky, quarrelsome, lazy, alcoholic, oversexed and prolific. In England too, it has been observed that: 'The stereotyping of coloured people . . . [is] based on generalisations from the less able and attractive to the whole population.'[11]

The 'ideal type' needs its 'anti-type', a compound of all the character traits opposite to those valued by the white Protestant. His puritan ethic demands achievement-orientation, honesty, cleanliness, loyalty, moderation in all things. His high-caste position in Ulster has been defined in a small booklet distributed by the Orange Order: 'When Loyalists have been considered as an element in Irish society as a whole they have been seen always, and have seen themselves, as a caste, analogous to the Hindu Kshatriya or the Japanese Samurai . . . The Anglo-Irish have the reputation of being the Prussians of England . . . The Ulsterman is . . . twice the man his southern neighbour is . . . '

A master-race concept? At least, to an Irish WASP, 'the Ulsterman has remained the prototypal Briton, patriotic, religious, industrious, mindful to the point of obstinacy of ancient fidelities . . . '

But, as for traditional Irish culture, ' . . . sham-Gaelic mystique and cult of heroes and martyrs . . . a tissue of barbarous piffle . . . a catalogue of demonology.'[12]

However, says the *Protestant Telegraph*, 'Ulster people are the earth's salt . . . the northern counties of Ireland and their people are very, very special indeed. The Ulster Protestant is a strong, robust character, with a fierce loyalty to his friends . . . His thoughts run only in "straight lines". He has a guileless innocence . . . with a built-in honesty . . . The Ulster Protestant has no time for double-dealing, shady dealing, hypocrisy and weakness. He despises traitors, political puppets, ecumenical jellyfish, opportunists, liars, crooks, apologists and snivellers . . .'[13]

And native Irish. The same paper's 'donkey-cart' stereotype is described in Chapter 8.

(Of all hated characteristics, the size of Catholic families seems to provoke Protestant hostility more than anything else. They resent the larger Social Security benefits paid to Catholic parents; but their main fear is that the Catholic birth-rate, 40% higher than their own, will result in their being outnumbered and in due course being voted into a united Ireland. Actually, the rate of Catholic emigration is double that of Protestants. The most recent surveys estimate that Catholics will not comprise a majority in Ulster before the year 2011 at the earliest, and perhaps never.[14,15])

The outcaste member, assigned to his subordinate role, serves the dominant group, more than all else, as a *scapegoat*.

Scapegoating is, and always has been, an intrinsically deeply satisfying process. The ancient Jewish faith called annually for a sacrificial goat on which the sins of the people were symbolically laid; the goat was then sent into the wilderness to die. The theme recurred in the Christian faith; forgiveness of sin followed the atoning death of Jesus. Nearly every civilisation, in fact, since history first began to be recorded, has made use of human and animal sacrifice to propitiate various gods and thus to ease communal guilt. And in normal daily life, as Gordon Allport points out, 'more than one person has complained to a frustrated friend after receiving an unwarranted attack from him, "Don't take it out on me." '[16]

When goals have been reached, it relieves guilt and enhances fallen self-esteem to be able to lay the blame, not on personal inadequacy, but on members of some minority group—either for occupying employment, or for contributing to community poverty by lacking it. It has been shown, for example, that American men whose job status had fallen were considerably more hostile towards Jews and Negroes than those whose status had risen;[17] also that anti-Semitism was far stronger in people who were not satisfied with their jobs than in those who were.[18] A great deal of experimental work, in fact, has established the important role of the scapegoat, or alternative target, as the object of displaced aggression following frustration.[19]

'Why are particular groups selected as scapegoats?' asks Professor L. Berkowitz. 'If all that is required for displacement is a non-feared victim, why are Jews attacked rather than, say, people of Scottish descent?'[20] Or, in Ulster, why Celts rather than Englishmen?

Berkowitz defines four 'stimulus characteristics' for the scapegoat —visibility, strangeness, prior dislike and proximity.

In the case of a Negro in a white culture, the visibility aspect is, of course, obvious. This feature in the Ulster Celt has already been discussed; I have tried to show that he too, like the Negro or Jew, provides the discriminating with multiple clues that enables him to be easily identified as a member of his group. (Remember, too, that 'visibility' refers to *all* the senses; it includes such clues as accent, name and address.) Discussing employment prospects for different immigrant groups in England, W.W. Daniel says: 'It is . . . impossible to escape the conclusion that the more different a person is in his physical characteristics, in his features, in the texture of his hair and the colour of his skin, the more discrimination he will face.'

Strangeness takes the process a stage further. The stranger—the Jew, the Black—is threatening to the prejudiced individual. The Catholic Celt, a member of a Church that values outward and visible signs, is thus easily perceived by the Protestant as having strange differences in culture and religion. The Catholic Church has a penchant for building vast walls round its institutions; the Ulster Protestant, with his talent for projectivity, depicts fearful rites taking place within.[21]

The concept of strangeness is heavily reinforced by metaphor—selected for outlandishness, generously mixed, and served warm to Loyalist readerships: 'The Roman Church is the Harlot Babylon . . . the Man of Sin . . .

the Ten-Horned Beast and the Son of Perdition . . . the Scarlet Woman.'[22]

The in-group, as mentioned earlier, sees the opposites of the qualities it values in the out-group and so has a prior dislike of this group. This in itself can be thought of as a subtle form of scapegoating, whereby the prejudiced individual projects on to the out-group member the qualities he most dislikes in himself—hypersexuality, for example. In a recent book, a psychiatrist tells of a consultation with an intelligent English mother whose daughter had decided to marry a coloured man. Although she was not prejudiced, the mother said, she was very concerned that her daughter's health might not stand up to it![23]

On what seems the same principle, the 'rabbit' metaphor appears to occur readily to any Ulster Loyalist putting pen to paper to complain about Catholic 'overpopulation'.[24]

Proximity of the scapegoat completes the list of the essential foundations of prejudice, and in Ulster is self-evident. By this token the Ulster Protestant blames his discontent on the Irish Celts rather than, say, the Vietcong. To paraphrase an old gibe—not only are they over-paid and over-sexed, they are over here.

'The Catholics have big families and they won't work. Daddy says we're poor because the Government takes his money to give to the Catholics and all their children and their priests to keep them in luxury because they don't work. In the South they wouldn't get any money and would have to work for it. Here they keep us all poor and don't appreciate what they get.' (Ulster boy, age eleven)

'They're against the Crown, but not against the half-crown.' (Well-known Protestant saying)

So the scapegoat is identified, but before it is finally driven into the wilderness the last doubt about the essential justice of its exclusion must be quelled. The in-group member, to defend himself from any consequent guilt, imposes *social distance*, a process whereby he symbolically places the victim at a distance well below his own on the evolutionary ladder. In its crudest form, this practice is very familiar: 'Dog—swine—pig!' Try, in fact, to think of one common animal whose name is not used as a term of abuse. And again, this tends to be applied to ethnic groups in certain special circumstances. The writer of *Apes and Angels* tells how, in Victorian caricature, the Irishman, as he began to compete for employment in England, degenerated from a harmless buffoon to a figure virtually indistinguishable from an orang-outang. Several of the monumental scientific follies of the Victorian era—the cult of physiognomy, Camper's facial angles, Mantegazza's morphological tree—reinforced the myth of Anglo-Saxon supremacy. By branding the Irishman as essentially different and basically undeserving, any guilt aroused by denying him employment was partly assuaged.

In Ulster, in the same way, expression of this myth can vary from the crude to the patronising. From 'the people of Ballymurphy are breeding like rats . . .', '. . . these estates are rabbit warrens' (letters to *Belfast Newsletter*), 'animals . . . sub-creatures . . . sub-humans' *(Loyalist News)* and 'I would starve them like rats' (schoolboy's essay),[25] to the subtleties of Terence O'Neill's speech (p. 38) and the over-denial of common expressions such as 'They're Catholics—they're very nice people' and 'I have some good Catholic friends'. (The *non-verbal* communication is, respectively, 'Catholics, generally, are not very nice people', and 'I am broad-minded . . . ')

With this central myth goes a cluster of others. The *myth of contentment*, for example, is often invoked when economic grievances are in question. Frequent dominant-group claims are, 'They *like* living in these crowded streets—it's so neighbourly . . . They wouldn't be happy in the suburbs without people to talk to . . . They moved them to nice houses and they kept coal in the bath.'

(This last dies hard. *Punch*, never notable for racial tolerance, published a cartoon in 1906 of a Jew nonplussed by finding a bath in the house he had just bought.[26] The coal-in-the-bath myth evolved shortly afterwards, and is still repeated with conviction.)

A striking experiment in an American primary school demonstrated the ease with which the most arbitrary myth can be accepted and acted upon. A class was told that blue-eyed children were superior in every way to those with eyes of other colours. The blue-eyes very quickly became aggressive, arrogant and punitive towards their classmates to an extent that not only provoked the latter to sullen and angry protest, but which also reduced the young teacher to a tearful wish that she had never started the experiment.[27]

Another self-fulfilling prophecy, perhaps? But what about protest? The in-group member, having accepted the myth of contentment, can only explain protest by blaming a small number of agitators for stirring up the people. This is generally the establishment line on any series of demonstrations or riots. The out-group *cannot* be discontented under the present regime, it is reasoned, via the myth of contentment; all disorder is attributed to a few trouble-makers or outside sources with hidden motives —Communism, Republicanism, 'international anarchy'. Dissidents are always discounted; the base of support for the regime, it is steadfastly believed, remains in the 'great silent majority' so beloved of conservative governments.

So the scapegoat is at last driven out; his pollution is inherent, and has been passed from father to offspring. No amount of wealth, no personal prestige can cleanse the outcaste stigma. Paisley's Ulster Constitution Defence Committee has as Rule No. 4: 'No one who has ever been a Roman Catholic is eligible for membership. Only those who have been born Protestants are eligible for membership.'

To maintain ritual purity, segregation becomes the rule—mainly in schooling housing, and marriage. There are '. . . crucial rules about marital and sexual behaviour, dieting rules. But also awareness of Cromwell's massacres at Drogheda and Wexford and the decisive Protestant victory at the Boyne are socialised into schoolchildren and they become as much a part of the consciousness of belonging to each faith as being black is for an American Negro. The fact that such aspects of group identity are invisible does not make them any less powerful in determining lines of co-operation or cleavage.'[28]

Mixed marriage, on the Protestant side, draws strong community disapprobation and often intimidation. On the part of the Catholic Church, the *ne temere* decree virtually precludes mixed marriage, anyway, by imposing complex and rigid conditions.

Discrimination soon becomes institutionalised; the growth of the Orange Order, a powerful vehicle for the maintenance of group supremacy, has already been traced. But the legend of Anglo-Saxon supremacy has for so long been told in the corridors of parliamentary government that the skill and subtlety of Ulster members in acting it out is consummate and probably unique.

Under the Local Government Act of 1922, for example, the Ministry of Home Affairs could alter electoral boundaries, effectively giving a disproportionate number of seats to Unionists (gerrymandering). In Derry, in 1969, the Unionist population (one-third of the total) held twelve seats on the City Council; non-Unionists (two-thirds) held only eight seats. There are similar patterns in Dungannon and Kilkeel where, in addition, housing is planned so as to concentrate the non-Unionist population into areas where they already have a majority. The gerrymander is thus a built-in regulator for ensuring Unionist control; it is probably inevitable, since the State of Northern Ireland itself owes its existence to a boundary which created an artificial majority.

Other grievances which provoked the protests of 1968 were the 'company vote', which allowed six votes to a man for every company he owned—and a voting system in which the right to vote in local elections was (and still is) limited to house-owners only. This operates directly against the Catholic large and 'extended' families and in itself disenfranchises 35% of the adult population. Without a job, of course, a man cannot buy a house. There are more detailed reports on boundaries and job discrimination elsewhere,[29-33] but a few figures are worth noting. For example, in Belfast's three largest firms, Catholics are employed in the proportions of 3%, 1.4% and 0% respectively. Further, the policy of attracting industrialists to Northern Ireland by provision of advance (sometimes free) factory space has allowed a concentration of industry in the predominantly Protestant East of the province and its diversion from the West. (In the first 20 years after the war, 455,580 sq. ft. of advance factory space was built in Derry (population 53,762). At the same time, in the largely Protestant towns of Larne (16,350), Ballymena (14,734) and Lurgan (17,700) the Government built respectively 777,700 sq. ft., 550,000 sq. ft., and 541,000 sq. ft.)

Once a concept of group supremacy has become part of the fabric of local and parliamentary government, one end-result, at least, is almost certain.

A ghetto forms when an ethnic group under common social and economic pressures concentrates in a segregated area to maintain group solidarity and for mutual protection. The growth of the Catholic ghettos of Derry and Belfast, for example, has followed exactly the same pattern as that of the Black ghettos of Harlem, Watts (Los Angeles) and Paradise Valley (Detroit).

In more detail, it works like this. Attracted by new industry, an agrarian population moves into the suburbs of an expanding city. To begin with, as latecomers they have the poorest ground and have to live at some distance from the industrial centre. (In Watts, travel to a factory in downtown Los Angeles can take two hours; in Belfast, the densely-populated Falls and Whiterock areas are several miles from the largest industrial complexes.) Besides this, it soon becomes clear that the members of the established group are not inclined to share their new-found prosperity and status and, especially where the new population's visibility as a group is high, they quickly attract discrimination and the scapegoat's role, by the processes outlined earlier. But the move to the cities continues, as farms become automated and as new, urban values are inculcated by television and newspapers.[34]

The new areas decline, very quickly becoming slums. The more able and adventurous members leave; the less able and mothers with families remain. At the beginning of this century, for example, Watts became a 'halfway house' for liberated slaves from the South. Wives, families and the unemployable remained in the area while fathers sought work elsewhere—much as is now the pattern in Strabane and Derry, with about one-third of bread-winners working in England.

The extended family is as constant a feature of the Deep South and of Latin America (where slums are at present expanding at an unprecedented rate) as it was of rural Ireland. In an agrarian community large families are valued, and, with virtually no mobility over several generations, children can multiply freely where relatives and neighbours are at hand to share in their upbringing.[35-39]

With urbanisation, the family is fragmented, and the adult–child ratio is disturbed. With an often-absent father and limited play-space, the youngster looks for a new group identity in the collective strength of the gang—which in itself becomes a *raison d'être* as opportunities for education and employ-ment dwindle. The term 'vigilante', by the way, came to Ulster directly from the Black ghettos, as did the cutting 'Uncle Tom', often applied here to a Catholic who joins the security forces or the Unionist party.

So if, as I suggested, the natural history of a ghetto is universal, how do pressures and conditions in the present-day Black ghettos in the United States compare with those of the urban enclaves of Belfast? (For the time being, we are not including political systems such as those of South Africa or Nazi Germany, in which a minority is forcibly segregated. This extreme stage has not yet been reached in Ulster, although planned housing segrega-tion and massive refugee movements plus physical barriers and internment have added up to a final result which is almost indistinguishable.)

Any physical differences between ghetto areas are largely in degree, and in most respects Ulster suffers heavily. Unemployment is endemic in any slum area, but the current estimated figure of 47% for fathers in the White-rock–Ballymurphy area (population 55,000) dwarfs Watts' 20% and 10-15% for Harlem and Black Chicago. Multiple occupation is not a common feature in Belfast Catholic areas; however, overcrowding is severe —especially now, when many families have been forced to squat in semi-derelict houses in the wake of the disturbances. Even before that, three or four children to a bed was not unusual, rooms were tiny and, if a child actually succeeded in getting a place at a grammar school, the stair-case was often the only place where he could have the necessary space and solitude to do his homework. And it was before the present riots, in 1967, that a survey by Building Design Partnership found 95% of the houses in both the Protestant and Catholic slum areas in Belfast unfit for human habitation, lacking a bath, a hand-basin, an indoor WC, and a hot-water supply.

Belfast has been spared at least a few of Harlem's problems—rat and cockroach infestation, a high intra-ghetto crime rate, and hard drug addiction. But for how long? There are now some reports of an increasing vermin problem in West Belfast, and gang warfare has escalated at a terrifying rate (see the next chapter). By 1972 in Belfast, gangs were younger and vigilantes older than in other similar slum areas (with Social Security benefits, there is less stimulus for a father to leave the area in search of work, and fewer older men do). While there was, at least up to about 1971, strong control of the younger element in the un-policed ghettos, with community leaders meting out rough justice (tarring and feathering, tying to lamp-posts), this control is now definitely breaking down.

Even as things stand, there is appreciably *less* fear and tension in the Black ghettos; a lively, bustling camaraderie contrasts sharply with the cold, watchful menace of the Bogside. One can enter and leave the area at will; there are no barricades or physical demarcation from White areas, and one is much less likely to be approached or molested. The Black ghettos, notably Harlem, can expand and are doing so—unlike the Catholic areas, bounded on all sides by barricades and barbed wire. More hopeful, less volatile, the Black ghettos are relatively free from the physical signs of present conflict —the burnt-out cars, the blackened hulks of buildings, and the marauding gangs of pre-teen children who make William Golding's *Lord of the Flies* look like a Sunday-school outing. The streets are lit in the Black ghettos, and there are cops on the beat. (In Belfast's Falls Road, I have several times recently seen a gang of boys, all in the 10 to 14 age-group, stop a car or lorry, order the occupants out, and then drive off, crashing or burning it. It is not unknown, on a cold day, for a vehicle to be hi-jacked and then burnt, while the youngsters stand round warming themselves.)

At this point one might ask how fair these comparisons are? Is it valid to freeze conditions at a single point in time, add and subtract physical differences and reach conclusions about, say, which ghetto the observer would live in if (God forbid) he had to choose?

A straight comparison is, I think, reasonable provided one major qualification is kept in mind. That is, that a ghetto has its own organic growth and development, and that one *could* be comparing areas that are at different stages. In other words, one should avoid the trap of drawing too many conclusions from parallels between an area in which certain advantages have been won at the expense of violent conflict and one only *now* passing through the phase of violence. Because, in the life-history of the ghetto, this phase is inevitable. Why? The answer, as it happens, is not patently obvious.

'Evils which are patiently endured when they seem inevitable become intolerable when once the idea of escape from them is suggested.'[40] Strangely enough, revolutions, revolts and protests do *not* typically take place against a background of economic hardship and political suppression. James C. Davies, reviewing antecedents of the major world revolutions, concludes that 'revolutions need both a period of rising expectations and a succeeding period when they are frustrated', and 'far from making people into revolutionaries, enduring poverty makes for concern with one's solitary self or solitary family at best and resignation or mute despair at worst'.[41] Revolution comes only when 'prolonged economic growth and political autonomy produce constantly rising expectations'. He gives as examples the French, American, Indian and Russian revolutions, and the race riots in the United States. There had been no revolution in Ireland under Gladstone or during the American Depression; it came *after* the period of extreme deprivation when, for a minority group, new expectations were frustrated.

Closer examination of the operation of this principle in a couple of present-day settings seems to bear it out. In the United States, from 1960 to 1967, the number of Blacks in the professional and technical groups rose by 80% (as compared with a 30% increase among Whites), and the number of Blacks below the poverty line dropped from 11 million in 1964 to about 8 million in 1967. But by the end of 1968, every major northern city had experienced a huge ghetto revolt.[42] In Ulster, violent protest has come with the long-delayed advent of post-war prosperity to the Province—or, rather, to parts of the Province.

And this precisely is the point. Earlier in this chapter I tried to establish that discrimination against a minority is most likely in an expanding economy. Now the idea can be taken a step further. It is *then*, when an expectation of reaching these new goals from a stake in industry is frustrated, that the protest business begins.

Probably the most constant and reliable sign of impending revolt is an increasing *consciousness of kind*—when the minority group, unable to expand outwards, turns back on itself, narcissistically, in a search for group identity.

'Black is beautiful.' A Negro no longer apologises for his race; hair-straightening barbers have gone out of business as Blacks have increasingly favoured natural, neo-African styles.[43] Native culture blossoms, just as in Catholic Ulster where the last ten years have seen a resurgence of traditional Irish drama and a trend away from the bucolic Ulster comedy favoured in the Protestant camp. More graphically, in early 1971, crowds of young boys roved through mixed areas painting 'I am a Catholic' on walls and pavements. Now there are names and pictures of traditional Irish heroes at every street corner, sometimes replacing the original street name and most youngsters in the Catholic ghetto areas have taken to wearing badges depicting earlier Irish revolutionaries; the most popular carries the profile of James Connolly. Similar in principle is the collective-ego-boosting material now common in Black districts, especially in school: glossy new books by Black writers, modern pictures and calendars depicting Martin Luther King, Jesse Jackson, Julian Bond. Several teachers in Black ghetto areas have told me about the children's new, aggressive pride in their race membership. Outside in the street are the names of the rather less reputable folk-heroes—Angela Davis, Malcolm X, Huey Newton.

To cement solidarity, blood ties are implicit in such group titles as 'Soul Brothers' (USA) and 'Irish Republican Brotherhood'.

Soon there are placards, marches, demonstrations. But, if protest comes with economic expansion, it brings in its wake another apparent anomaly: while initial protest by the minority is usually peaceful, aggression, at this stage, comes from the *majority* group. Since, on the face of it, a deprived minority seems to be, of the two, the one group more entitled to anger, why this apparent reversal of reaction?

Here is one of the odder quirks of human behaviour. 'Most of us when we do a caddish thing harbour resentment against the person we have done it to, but Roy's heart, always in the right place, never permitted him such pettiness. He could use a man very shabbily without afterwards bearing him the slightest ill-will.' (Somerset Maugham, *Cakes and Ale*)

This, the paragraph suggests, was unusual. The same human trait has been described more scientifically, if not more elegantly, as cognitive dissonance.[44] In the field of protest, it operates like this:

As the minority uses various means to draw attention to its plight, the in-group member becomes more aware that he has a just cause for grievance. But this makes him no more willing to share his privileged position; his conscience reproaches him as a result and an internal struggle follows. As the prejudiced individual has, anyway, a tendency to solve his conflicts by projection, he loses no time in turning his anger outwards on its perceived cause —the out-group member. Harbouring fantasies of riddance of the group whose presence disturbs his equanimity, he will act them out at the first opportunity. He develops a heightened readiness to react to cues to aggression, and the slightest provocation—a flag, a song—can result in days of fierce rioting (as, for example, did the display of an Irish flag in a Belfast shop window in 1965).

This mechanism probably underlies the pattern where, in the Ulster situation as in race and campus riots elsewhere, physical violence usually has its first expression in a display of disproportionate *force majeur* by members of the establishment (security forces, Orangemen) directed towards the out-group members.[45] The events of Derry and Burntollet in 1969-70 are our classic examples. Violence from the *minority* is a reaction, and comes later.

The progress of events from 1969 onwards, the interplay of aggression and counter-aggression, has been carefully traced by other writers, and is beyond the purpose of this book. But a couple of quotations are worth looking at now. Nielsen, for example, after a penetrating study, concludes: 'From absolute deprivation there arises an awakening that starts a general model of development. Where aspirations for development are blocked the development model is broken off and the society goes into a conflict phase. Political leaders are not able to handle the conflict, the Army fails to stop the conflict spiral, and the process is brought into what can be described as 'a second conflict spiral with a third party (the Army) included . . .

'And until something or someone is able to break the vicious circle it is . . . most likely that the violence will continue, and maybe even grow worse in Ulster.'[46]

Lumsden, who has published extensive studies on the conflict in Cyprus, also stressed the unacceptability of a third force to both parties in a conflict and recently applied this finding to the role of the Army in Ulster.[47,48]

Davies concludes his review of theories of revolution by proposing ' the utter improbability of a revolution occurring in a society where there is the continued, unimpeded opportunity to satisfy new needs, new hopes, new expectations . . . It is of course true that in a society less regimented than concentration camps, the rise of expectations can be frustrated successfully . . . This, however, requires the uninhibited exercise of brute force as it was used in suppressing the Hungarian rebellion of 1956. Failing the continued ability and persistent will of a ruling power to use such force, there appears to be no sure way to avoid revolution short of an effective . . . continued response on the parts of established goverments to . . . the needs of the governed.'

Davies, by presenting a frank either–or alternative, risks over-simplifying. But we have, so far, been dealing largely in generalities. The next step must be to focus attention more closely on a special section of the community—its children. Is it the fear that necessitates living in sectarian purdah—translated into aggression—that has created Ulster's growing youth problem? Is it a typical, or a unique, ghetto revolt?

8

The Route from Fantasy

David is twelve. He said, 'The favourite game in our street is "riots". We use tomato sauce for blood. My friend went in to his mum with his face covered with tomato sauce and said he had been hit with a bullet. When she found out it was just sauce she was furious and kept him in.' He laughed.

When I saw him a week later, David had a huge bruise on his forehead; his left eye was swollen and badly cut. This time it had been a real riot, with real blood.

Another twelve-year-old likes playing 'rioters and soldiers' during his school lunch-break.

'God help the unpopular boys,' he said, 'they are always the soldiers.' In the evenings, he and his gang go out armed with stones and petrol-bombs, hoping for an unwary Paratrooper.

Cause and effect? More often now in Ulster, the boundary between childhood fantasy and harsh reality blurs. The boy in the soldier suit with his toy gun, ducking into a doorway, firing, running bent double, firing again, may have seen yesterday a soldier do exactly the same thing—and not as a game. Rich material for fantasy is available in a way it can rarely have been before, even as conveyed by the most skilful Method actor on TV or in the cinema.

Given a few facts and a few more anecdotes, it is temptingly easy to propose a simple cause-and-effect mechanism. Riot games in the back yard. *Ergo*, stones and petrol-bombs in the street. But is it *too* easy? How justified is the distinct unease conveyed with reports that Belfast children are given toy guns at Christmas, and that they enjoy playing riot games.[1] In March 1971, for example, a report from the National Society for the Prevention of

Cruelty to Children read: 'In the playgroups . . . some of the children between the ages of four and five (particularly the boys) spend considerable
time erecting barricades across the floor, pretending to shoot and throw
petrol-bombs.'[2]

This brief paragraph received national coverage and attracted a long jere-
miad from at least one London daily. I can still remember the dawn trek to
Broadcasting House on a drizzly Belfast morning to 'answer' for these children on the Today programme. It was natural, I said then, that children's
games would reproduce what they saw happening in the adult world (and
whether or not these events were to be deplored is not the point at issue).
Put in another way, it would be more astonishing if children living in a riot-
torn area did *not* play riot games. In the search for stresses and stimuli that
provoke children to habitual street violence, we have to look for more than
the possible effects of learning in unstructured play.

This question also crops up very often: should children be specifically *encouraged* to play aggressive games? Are aggressions acted out in the sandpit
less likely to be acted out in the streets? This is, of course, a vitally important question for anyone who has to deal with children, particularly in areas
where there is a high predisposition to violence; it is a question which must
be answered.

Theories of aggression are legion, and a critical review is beyond the
scope of this book. Most, however, break down where it is attempted to fit a
single theory to all conceivable situations. For example, the well-known
'frustration–aggression' hypothesis, first formulated at Yale in 1939, proposed that all aggressive behaviour was the result of previous frustration
and also, that 'the existence of frustration always leads to some form of
aggression'.[3]

This looked at once bold and simple. But other researchers have found
since then, as the theory was tested in various settings, that they have had to
add one qualification after another. Having now been modified several
times, the current and most widely-accepted form of the theory is broadly as
stated by Berkowitz: 'a frustrating event increases the probability that the
thwarted organism will act aggressively soon afterwards. This relationship
exists in many animal species, including man.'[4]

This, of course, says very much less than the original statement. The
trouble is that these 'umbrella' theories of aggression are (like umbrellas)
cumbersome, potentially dangerous, and basically inefficient. A working
theoretical framework, evolved from detailed observation in a given situation, is, I suggest, by far the most useful tool for dealing with that same
situation.

Where violent gang behaviour in the Ulster setting is concerned, I suggest a simple three-stage sequence, one that could be applied in most areas where ghetto youth finds itself in conflict with authority. First, the child's 'anti-hero' or bogeyman is defined; fear is aroused to the point where the child is obsessed with fantasies of riddance; then, given specific conditions and cues, these fantasies are ultimately translated into action.

Freud defined three components of human personality—the id, the ego, and the super-ego. The id comprises basic 'instincts', the primitive drives to seek pleasure, to kill or destroy. To a child brought up in Western civilisation, this is the 'bad part' of himself. The super-ego is his conscience, a set of learnt standards and ideals for which he must strive. The ego is 'me', the man in the middle. I have already discussed, in Chapter 5, the way in which these three components have throughout history been projected on to the Devil, God, and Man respectively. Centuries before Freud, Paul of Tarsus, in a moment of stark insight, wrote: 'For I delight in the law of God . . . but I see another law in my members, warring . . . bringing me into captivity to the law of sin . . . O wretched man that I am! . . . With the mind I myself [ego] serve the law of God [super-ego] but with the flesh [id] the law of sin.'[5]

Even after so long, the mental anguish still comes through. And a modern child's internal conflict between the 'good' and 'bad' parts of himself can be just as cruel. He too must project. With the gradual demise of the old and simple Sunday-school concepts, other figures begin to replace angels and devils. So games of cops and robbers, Cowboys and Indians, British and Germans, increase in popularity; in the cinema, the detective and the criminal; in pantomime, the Fairy Godmother and the Demon King. There must be goodies and baddies—and the goodies must win. From this, the happy ending, the child gets great satisfaction. He is *resolving internal conflicts by projection.*

Now these games, as well as being satisfying, are safe. There are no Indians, nor is there a Demon King, in the next street. But what if, by combination of circumstances, the image of 'baddie' or anti-hero becomes projected on to an available group in the child's own community? And what then if a further set of circumstances favours active aggression against this anti-hero?

In this chapter, I want to suggest that this is precisely what has happened in Ulster, that the bogeyman to the Protestant child is the Catholic, to the Catholic child the British soldier, and that the train of events of the past few years has specifically favoured violent and uncontrolled action against him.

As a starting-point, let me take up again a point made in Chapter 2—the importance of the break-up of the extended family, the impact of urbanisation on the traditional, large family groupings. With fewer relatives to share in child-rearing and a frequently absent father, the child–adult ratio is exceptionally high, and the practical results are a lack of supervision and loss of family group identity. The child, looking for a new group, finds the gang. In Belfast, in fact, gangs are bigger, more numerous and younger than anywhere else; a further unique feature is that they tend to be approved, at least tacitly, by adults, who look on them as a first line of area defence. And, in the gang, there are new behavioural norms. Both major religions in Ulster are traditionally authoritarian and, as controls weaken, anti-authority behaviour tends to be approved by the gang and to make for status within it.[6]

It was into this potentially explosive situation that a terrifying cultural anti-hero was injected. 'Terrifying' is no overstatement. Most people, I find, greatly underestimate the role played by fear in events behind the Ulster barricades. On the young Catholic, a British Tommy has much the same effect as a steel-grey stormtrooper with a swastika on his helmet and Nazi jackboots would have on an English or Jewish child. And this even before the first rubber bullet is fired.

True, the recurring situation in itself is horrific—a scene from 'War of the Worlds'. After a giant bulldozer has burst open a barricade with the force of a bomb, the sight of a line of masked soldiers emerging through the smoke, guns trained, has certainly made at least my scalp tingle. And I have seen women turn chalk-white on identifying a shout in the street as 'an English voice'.

But the fear goes further back in time than these, surface, events of recent years—deep into Irish culture, in fact. Shortly after the British Army entered Belfast—*before* they were engaged in conflict in Catholic areas —Sean, aged twelve, produced a picture portraying himself as a warrior in traditional Irish garb, and underneath wrote this poem expressing his wish to drive out the British invader (Wolf Tone in 1795 founded the United Irishmen, predecessor of the Irish Republican Brotherhood, later the IRA):

> We go to fight the Crown
> To drive them from our land
> Why should we be killed and beaten
> We are the freemen of Tone's band.

Anti-Army attitudes among children were, in fact, common long before the period of open warfare between themselves and the troops.

With Army units active in Catholic areas, the fear component emerges much more strongly. It was Sean, six months later, who drew two remarkable cartoons: The first is of a soldier, a bovine half-devil, half-human, sprouting horns and a tail; a trail of destruction lies behind him. He wears a kilt: Scottish soldiers, believed to come from strongly Protestant areas, were held in particular fear by Catholic children. The other drawing is a bitter satire on Sean's environment, entitled Tir-na-Og, Irish for 'The Land of Children'. A shadowy corpse and burning houses in the background are dominated by a soldier with semi-Eurasian features, sketched in hard, cruel lines.

While Sean's facility for reducing his situation to a vivid proposition in black and white is certainly unusual, this general theme, casting the soldier in the role of demon marauder, is a quite typical response from Catholic schoolchildren. Two of many drawings in my possession, all done spontaneously, are entitled 'British soldier shooting children'.[7]

It is perfectly true that some children have been killed recently by Army gunfire. But these drawings were done at least a year before the first child had died. So why this consistent image of the terrifying anti-hero?

> What shall we do with the Para bastards,
> Early in the morning?

> Kill, shoot, burn the soldiers (repeat)
> Kill, shoot, burn the bastards,
> Early in the morning.

> . . . .

> If you hate the British soldiers, clap your hands
> (clap) . . . etc.

The soldier, in fact, started with a huge disadvantage. It has already been suggested that a simple frustration–aggression hypothesis, although applied in the case of adults, is not so directly apposite to that of young children who are not nearly so aware of social or economic frustration.

The *adult* Celt is frustrated by Anglo-Saxon domination—now backed by armed might and personified in the British soldier. It is the adult who, by his attitudes to the British link, has cast the soldier and the Protestant in the villain's role, and who has reinforced this by a rigorous process of indoctrination beginning from birth. To the Protestant, on the other hand, the economic and social frustrator is of course the Catholic in the community—so the twin processes of education for bigotry are almost exact mirror-images of one another. In short—the adult's perceived frustrator is the child's anti-hero. How is this doctrine inculcated? On the Catholic side, rigid insistence on separate schools goes hand in hand with socialisation into a strongly Irish-national culture. But, nationalism apart, evidence suggests that segregation in itself is productive of group antagonisms between different ethnic groups. In a boarding-school exclusively for Jewish boys, for example, boys were found to express more anti-Christian attitudes than did Jewish boys in mixed schools—although the school believed that its teaching was angled towards tolerance, and that nothing in its environment explicitly fostered anti-Christian feeling. Many of these boys may, of course, have reflected attitudes of parents who preferred exclusively Jewish schooling; at the same time, the effects of 'teaching tolerance' in a segregated environment are as minimal as one might have expected. In a further experiment at the same school, in which one dormitory was told stories designed to create a 'consciousness of kind', the boys in the dormitory became increasingly aggressive towards other groups.[8]

Conversely, 'equal status contacts' have been found to *lower* prejudice. In a study of the US Army during World War II, anti-Negro feeling was found to be lowest in mixed Companies[9]—as also among housewives living in inter-racial housing projects.[10] The same kind of attitudinal change has been seen among boys at mixed racial camps,[11] and among children who have transferred to integrated schools.[12]

These results of segregation are not surprising—experimental findings simply support the surface probabilities. A group of children forbidden to mix with another group even for lessons or games can hardly fail to perceive the others as in some way strange and threatening—and, indeed, to be seen in the same light themselves. This must apply where the segregated schools in Ulster are concerned, where this consciousness of kind permeates every part of school life—lessons, drama, games. Although both Catholic and Protestant Church leaders, on whose insistence this system still exists, deny its importance as a factor in the conflict over Ulster, Lundberg and Dickson, reviewing work on school segregation by ethnic grouping, conclude: 'it is hopelessly contradictory for a group to want to maintain a distinctive group identity but to expect not to be discriminated against on the basis of this exclusive identity.'[13]

And it is *ethnic*, not religious, differences that are under discussion, in the school as well as in the adult world. A few examples of group pressures brought to bear on individuals may help to make this clear.

Michael, a fourteen-year-old Catholic, moved to Belfast two years ago from his school in England, where he was a keen footballer. When I asked him whether he had been able to keep up his football he shook his head sadly.

'The only games we can play in our school are hurley and Gaelic football. I once told the Brother that we wanted to pay soccer and rugby. He was very cross; he said these were British games. He went to see my mother about me.'

Jim, thirteen, a Protestant, said:

'*Me* go to a hurley match? That's a Fenian game. *I* know what they can do with their Fenian hurley-sticks . . .'

It was as long ago as 1968, in terms of events, when a Catholic boy told me: 'We are doing a play in school about the Irish rising. I was the leader; we wore black [IRA] shirts and berets.. The British soldiers are slimy and can't be trusted; we always get the better of them. I like the play.'

Children from four other Catholic schools have told me about their school drama, and the Irish rising of 1916 seems to be a favourite subject. Invariably the IRA soldier is the hero, the British Tommy the villain. It was four years before a British soldier appeared on the streets, and five years before the IRA emerged as a modern fighting force, when the boy I have just quoted showed me his black shirt and beret. He had never seen the Englishman in khaki, but his bogeyman image was even then deeply placed in the Catholic consciousness. Bernadette Devlin, in her autobiography, graphically recreates the anger of her headmistress when she came to inspect a class history project and '. . . old Horatio Nelson caught her eye. In one blinding flash she realised that her patriotic Fenian wall was decorated with British generals and British heroes, and she just tore the chart, from one end to the other, right off the wall. She crumpled it up, stamped on it . . . She was white . . .'

From the Protestant side, Joe, eleven, said: 'At school they try to get you into the Junior Orange, and once you're in it's almost impossible to get out. They call you a Fenian-lover if you speak to a Catholic boy and will beat you up. The Catholic school up the road lets them go to school without uniforms so as our lot won't pick them out. But *we* know them.'

Jeremy (twelve): 'The boys in school are against Catholics. They sing songs about the Pope and all. If you won't join in or go to the Lodge they call you a Fenian-lover.

'I went out to play with Seamus the other night. We met a man who was drunk and we spoke to him. He bought us a Coke each because he said he had never seen a Catholic and Protestant boy playing together before. He said it was great. He was very drunk . . .'

At extremes—remember Anne's hallucination (p. 74)? The really telling thing about the apparition, which she drew for me, was her reason for making the man a Protestant 'because he was dressed in funny clothes, and was different from us'. Certainly the figure, in strange checks and colours, is outlandish in appearance, and illustrates dramatically the almost total lack of contact between the two child communities.

Two other drawings offer equally revealing commentaries. Two unrelated children who live in Unity Flats (the Catholic block in the Protestant Shankill Road) drew, at different times, pictures of their home. The area around the flats has often been the scene of fierce fighting. Intriguingly, each child drew a mediaeval fortress with no physical resemblance to the reality. Do they see themselves as beleaguered, as among enemies? If, as I suggest, they do, then these two crude drawings condense the perceived plight of the Catholic child in Ulster into a brief, poignant message that several thousand words might fail to convey.

As consciousness of kind strengthens, it tends to colour all contacts between members of the opposite groups. In any quarrel, undesirable traits are attributed, ultimately, to group membership.

Bernadette (eleven) said: 'I only met a Protestant once—a wee boy in hospital. The first time I spoke to him was one night when I wanted to see "Top of the Pops" on TV and he wanted to see a cowboy film. I wouldn't turn over, so he said it was the Catholics that had started the riots. He said you're a Fenian . . . then a bad word. I told Sister on him, because I knew it was the Protestants started the riots; I seen it with my own eyes.'

At eleven, Bernadette was puzzled by a reaction whose vehemence she ill understood. Thirteen-year-old Kenneth, on the other hand, told his story with unconcealed glee. At the time (1969) he lived on a 'mixed' estate—now wholly Protestant.

'The old woman next door to us is a Catholic. On Saturday we pulled leeks out of her garden and she complained to my granny. They had a smashing row over the fence. In the end my granny said, "Get back up the Falls where you belong, you old Fenian bitch," and *she* said, "Get back up the Shankill, where *you* belong." When my granny came in she was raging. She said, "You can pull whatever you like out of her fucking garden; *I* don't care." '

Thus, in the event of conflict or competition, is the stereotype at once invoked and strengthened.

Confronted with an ogre, the child reacts initially much as one might expect . . .

If it was not so tragic, a touch of ironic humour would be perceptible in a clinical situation that confronts me from time to time; I have to admit to the same psychiatric ward Catholic and Protestant children who are emotionally crippled by precisely similar fears, each of the other. The only differences in the types of fears expressed are age differences. In general, children under the age of puberty are obsessed with the possibilities of physical violence; their fears are that the 'others' will burn their houses, shoot them or their parents. Almost always, as pointed out in Chapter 6, these forebodings reflect what is expressed at home by the children's parents. Older children are more likely to mention social and economic risks: on the one hand, fear of Catholic population increase coupled with resentment of Social Security benefits given to the larger Catholic families; on the other, fear of continued Protestant job domination.

These are, of course, real possibilities, but are they in themselves the basis for these children's fears and hostility? The main clue is to be found in the quotations and drawings I describe in this chapter; they are based, just like the adult attitudes, on group stereotypes that have been part of the two cultures for at least a hundred and fifty years. The dual image of the Irish Celt in Victorian caricature—the feckless peasant and the simian terrorist with bomb and gun—still have their precise counterparts in Protestant publications. For example, in a cartoon in the *Protestant Telegraph*, a Southern Irish 'plain-clothes' policeman (i.e. in a donkey-cart, with the stock-in-trade battered hat and clay pipe) reins in to congratulate a colleague who has just been mobilised as a one-man 'flying column'—on a bicycle. The *Loyalist News* runs a strip cartoon, 'Bill and Ben, the IRA men'. The *dramatis personae* are virtual replicas of those in Tenniel's sketches of 1860-80—armed, undersized, ape-like, scar-faced and sinister.

Sean's drawings are equally vivid reincarnations of the hard-faced, helmeted British soldiers, tramping through the smoke of burning villages, common in Irish magazines of the late nineteenth century.[14]

What I am suggesting, in short, is that these children's fears are *not* reality-based—that is, linked to events within the child's own experience. He does not fear, primarily, the flesh-and-blood British soldier or the marauding Protestant, but his bogeyman equivalent—the old cultural anti-hero.

This, at first blush, might appear so much nonsense. Protestants *have* burned out hundreds of Catholic families, English soldiers *have* wreaked havoc in Catholic areas. It is natural that the youngsters are frightened. So why this suggestion of the 'cultural anti-hero'?

It is basically a question of timing—of what comes first. A stereotype (Chapter 7 is born in a situation of economic competition—when there are not enough jobs to go round. Exclusion of the minority group from employment is rationalised by attributing to its members undesirable qualities. Frustrated by this discrimination, the out-group finally takes to open demonstration against the majority, who then point to this behaviour as illustrative of the out-group's essential unworthiness. The justice of their policy of exclusion is thus reinforced, and still more rigorous measures are instituted. This, the perfect example of a self-fulfilling prophecy, has been enacted in miniature in the classroom (p. 104); but also on a full scale in almost every plural society in the world. First the stereotype—*then* the undesirable behaviour.

So experience and fear do not exist in a simple cause-and-effect relationship; the complete reverse is often true. This seems to be the case at any rate where Ulster ghetto children are concerned: most of the drawings and quotations expressive of fear and hatred *pre-date* the period of street violence.

How far do they reflect the factors that actually precipitated the violence?

The *Loyalist News* has a short way with dissidents: 'John Hume told a mass rally that some of his constituents who have been interned are to go on hunger strike. This is the best news we have had for a long time. Just think of over 300 rebel animal funerals . . . We Loyalists wish them a long and steady fast, and we only hope that they won't give up. So don't disappoint us, you rebel scum.'

And, ' "I am an outlaw on the run." So says Paddy Kennedy, Republican Labour MP for Belfast Central. We have only to say, if you are seen in Belfast before the Security forces see you, don't expect to be returning to the South. Our motto is: shoot first and ask questions afterwards.'[16]

In short, as a clergyman writing in the *Ulster Protestant* notes approvingly: 'With all these recent outrages . . . more bitterness and anti-Catholic feeling has been generated than ever before . . . I thank the Provisional IRA. God will use it for good.'[17]

A strongly-felt fear, in the adult as well as in the child, is closely shadowed by its corresponding fantasy of riddance. In the competitive situation, each group muses on the benefits that would accrue to itself if the opposing group were simply eliminated. The child sees, in particular, release from his recurring fears. In the Ulster Protestant camp, the theme of physical violence has begun to dominate Orange songs and right-wing speeches since, over the past few years, more subtle methods of eliminating the opposition (such as forced emigration following discrimination in housing and jobs) have become progressively outlawed by legislation.

'I don't want to get the rebels off the streets. I want to get them under the ground,' said the Revd Ian Paisley in Ballymena to wild applause.[18] More recently, a hundred thousand Ulster Vanguard supporters roared approval of Mr William Craig's announcement that he was keeping a list of persons to be 'liquidated' at some unspecified time. And, of course, to dangle a favourite fantasy before any audience or readership will sell political publications as certainly as it will sell pornography. They will buy it and love it —although they do not necessarily become more likely to *act out* the fantasy. But I anticipate.

On the other side of the coin fantasies of riddance, this time from the Army, are as evident in the Celtic population. Two favourite, and characteristic, songs have already been quoted, and the wave of sprawling wall graffiti continues: 'Army out . . . kill . . . burn . . . shoot.'

'We will not rest,' said a Provisional leader, 'until the last British soldier has been driven off Irish soil.'

So, for the child too, the anti-hero must be driven out or destroyed. This means physical force; the concepts of economic or geographical sanctions are practically meaningless to a pre-pubertal youngster.

Billy (ten) says, sweepingly, 'All Catholics should be killed or burned.' The first hint of awareness of more devious methods comes later . . .

'I would make things so bad for RCs that they would be rushing to the border to get out of Northern Ireland. One of the first things, I would close all the labour exchanges and social security offices so they would not get any money, because the way I look at it is, this is a Protestant country so the Catholics should not be in it, they should be thankful and not complaining.' (Fifteen-year-old boy, quoted in *Newsweek*, 19 April 1971)

The same article includes the paired, typically direct responses of two younger children:

John, twelve, Catholic: '. . . I would abolish the border, get rid of the Unionist Party and shoot Paisley.'

Alan, thirteen, Protestant: '. . . I would have Bernadette Devlin shot for treason.'

A Protestant boy, aged ten, draws his street. On either side are the houses, and in between are himself and his friend. Splashes of vivid red run between the pavements—'Fenian blood'. This is the day, he says, that he is looking forward to.

George (eleven), also a Protestant, draws himself and a friend attacking Unity Flats, helped by the Army.

Gerard, a twelve-year-old Catholic, paints a soldier hoisting the Republican Flag, dripping blood, before a blazing House of Commons; a torn Union Jack smoulders at the soldier's feet: the soldier is Gerard himself.

Terrifying fantasies, it is true. Will George and Gerard act them out?

'In this supreme hour the Irish nation must, by the readiness of its children to sacrifice themselves for the common good, prove itself worthy of the august destiny to which it is called.'

'Dear Ireland, take him to thy breast this soldier who died for thee; within thy bosom let him rest among thy martyrs sanctified.'

> (Death notices of a thirteen-year-old
> member of the Junior IRA, shot by the
> British Army)

The bogeyman has been identified, and the original fears have been intensified by the self-fulfilment of age-old prophecies on the streets for all to see. Now, under what conditions are fantasies of riddance translated into action? These several major conditions were fulfilled in Ulster by 1970. In approximate order of importance, they were:

Proximity: this may be self evident; still, all factors should be considered in the search for alternatives to violence. If Nazi stormtroopers had entered London streets, English children might, or might not, have thrown at them every object to hand. But they had to be there. And, since they never made it, the armies of the Third Reich were at least safe from children's stone-throwing. Similarly, in Ulster, in a central militarised zone between mutually hostile areas, the British Army would be much less likely to draw mob attacks than they are by their presence in residential areas.

Modelling: a child is, all the time, looking for behavioural cues from adults. A very large literature establishes the prime importance of imitation as a determinant of aggressive behaviour in children. For example, children witnessing adult aggression have been found to be much more likely to act aggressively themselves soon afterwards,[19,20] and this style of behaviour can be evident more than six months later.[21] Not only that, but an adult, effectively condoning aggressive play by turning a blind eye, can increase the likelihood of future aggression.[22-24].

On the streets, the pattern of adult–child imitation is clear to anyone who has watched a racial or political riot. 'The big fellows were throwing stones, so I just joined in,' youngsters often tell me. Partly, this is a shifting of responsibility, but the cue, if not the motivation, came initially from the adults. Several Belfast mothers, too, have told me that their young son's 'hero' is Joe Cahill, or Dutch Doherty, the IRA leaders.

(The outlaw, from Robin Hood to Two-Gun Tex, has always held an intense glamour for any youngster. As a figure on to whom he can project undesirable wishes and anti-authority attitudes, the outlaw serves as a valuable release from anxiety, but, as a behavioural model, he is a disaster. The latest myth about Mr Dutch Doherty is that he shot up a pub TV set after the horse he had backed lost a race. The barman objected. 'Whatya gonna do about it, then?' enquired our hero, twirling his revolver. Legend has not yet added that he strode off into the sunset . . .)

On a political level, the Illinois White House Conference on Children and Youth, in 1970, concluded that major causes of youthful violence were: 'The Vietnam war: violence endorsed, no matter with how much regret by men of stature, brings violence into the area of acceptable behavior;' and 'Provocation by authorities: personnel in law enforcement organisations, correctional agencies and schools tend to over-use physical methods of restraint or discipline . . .'[25]

Modelling on *verbal* agression is just as likely. The violent phrase from a verbally agile politician or preacher can have its direct counterpart in a violent *act* on the part of a man (or child) without the capacity or education to ventilate fully his aggression verbally. He uses the only way left—physical force. The first man, with manifest injustice, then 'condemns' the second. But he has himself been providing conflicting cues. Each demand for, say, rearmament of the police provides another cue to aggression.

So do guns themselves. One experimenter, for example, showed that a man who had been frustrated in some way was likely to behave in a much more aggressive manner to his frustrator in a room where a gun was visible.[26] proceeding from this example, Professor L. Berkowitz has drawn attention to other 'aggression-drawing stimuli.'[27] Flags are common examples. The display of a tricolour in a Belfast street in 1965 sparked off three days of fierce rioting; conversely, nothing excites the fury of a Catholic crowd like a Union Jack. As polarisation increases, so does hypersensitivity to these cues, often to the point of near-paranoia. In Derry in 1969, after an attack by Civil Rights marchers on a police station, Mr John Hume (Opposition MP and a leader of the demonstration) blamed the authorities who had re-routed the march past 'this provocative building'.

And, from the *Protestant Telegraph* of 21 August 1971: ' . . . waitresses in the Belfast Europa Hotel are sporting green jackets, white blouses, and orange skirts. Such a combination of colours should be sufficient to deter Loyalists from using the facilities of that establishment . . . The Chairman of the Community Relations Commission and John Hume (SDLP) have frequented the Europa. Does this colour scheme meet with their approval?' Touché?

The Family: We have seen that, where the extended family is split into nuclear groups, a disturbed adult–child ratio, with resultant lack of supervision and growth of gangs, is a natural consequence. One additional point is worth making here: that is, that most mothers in the ghetto areas, contrary to impressions given in the Press, do make strenuous efforts to prevent their children from becoming involved in street rioting and gang warfare. But a mother is often in conflict with older children and with her husband —and it is physically impossible to keep indoors a large family who are determined to go out. The concept of modelling is again marginally relevant; the child reads non-intervention, for whatever reason, as tacit approval.

Frustration: We have already looked, briefly, at the frustration–aggression hypothesis in relation to the Ulster disturbances. I have suggested that, while frustration is the major factor provoking aggressive behaviour in the adult and adolescent population, fear of a proximal anti-hero plus modelling are much more important factors where children are concerned. But for the youngster there is frustration too; he reacts less to the primary frustrators (poor housing, discrimination, etc) than to the *secondary* frustrators—those that have resulted from the conflict itself. Anger is often expressed by children at occupation of playgrounds and schools by troops, repeated detention and questioning, limitation of movement, dawn raids and internment of relatives.

Incidentally, the *fear* that a parent may be interned is much more likely to cause anger and anxiety than the *fait accompli* itself. This event, when it comes, is often greeted with curiously mixed feelings. A certain status in the street and in school follows the father's internment, and harassment by the Army is reduced. Also, it is reasoned, he will not, at least, be blown up or shot in the street. In my experience, sons and nephews of internees tend to be leaders of militant groups. This may, of course, be due to inherited characteristics rather than acquired status, but the sequence of events usually suggests that the second factor is the more important. Boys tend to join these groups or rise to leadership after, rather than before, the male relative's arrest. In a culture where sons and nephews are numbered by the dozen, a policy of internment can only be wildly counter-productive.

Any case for the importance of *primary* social frustrators in provoking youthful violence is fairly unconvincing. For example, the Illinois Commission on Children reports as a major cause of violence: 'crowded living conditions: people living in crowded conditions could be expected to produce an increase of violent interaction.'[25]

Perhaps—but the case is hardly proven. Deplorable as some youngsters' living conditions may be, the process by which discontent erupts into violence is, as I have suggested, much less direct than this. The myth of seasonality, too, crops up from time to time; the long, hot summer is cited as a cause of urban riots, and it is quite easy for the reader to imagine frustration and rage rising with temperature. Anger and heat are, of course, closely related concepts. But the implication, again, is too ingenuous. It is true that riots are more common in the summer months, but the extra free time and boredom that accompany the long school holidays are probably the main factors favouring disorder.

Exploitation. During the last phase of the Congo rebellion, in 1964–5, Simba terrorists were increasingly thrown back on the 'human shield' tactic. Touring villages in lorries, they attracted young boys with stories of adventure, patriotism and money, then sent them over in waves, with guns and sticks, to attack the government army. Hundreds were killed. When most of the troops' ammunition had been exhausted, the adult terrorists emerged from hiding and launched their main attack. Now, travellers report some Congo villages almost empty of boys between the ages of 13 and 19.

'Women . . . screened a man who had shot at soldiers when troops returned his fire . . . pushed their children out into the street.' (*Belfast Newsletter*, 5 January 1972)

This episode was slightly unusual in that the youngsters had to be actually pushed into the front line; in general, stone-throwing children congregate anyway at the front of rioting crowds. This natural tendency is heavily exploited in the belief that the Army will hesitate to open fire. It is a very dangerous belief, since gunmen have often used such crowds as a cover; several children have already been killed in cross-fire and many more injured.

The junior terrorist organisations, of course, represent *planned* exploitation on a very large scale. General Grivas, a lifelong admirer of the IRA, has commented, 'I know of no other movement, organisation or army that has so actively employed boys and girls of school age in the front line. And yet there is every reason to do so: young people love danger; they must take risks to prove their worth.'[28]

This is of course a piece of quite unscrupulous rationalisation. It represents deliberate confusion of 'love of danger' with what is, simply, a relative failure on a child's part to realise that the danger exists at all. In Belfast I have seen children run up to within a few yards of a soldier with an aimed high-velocity rifle and lob a petrol-bomb over the sandbags with a nonchalance that no adult could imitate. In part, the youngster, like his elders, believes the soldiers will not shoot at him, and doesn't consider the possibility of cross-fire. But more than this, the pre-adolescent child, with undeveloped death-concepts, is not psychologically capable of weighing up, in realistic terms, all the possible consequences of his action.[29] These limitations are absolutes, virtually independent of environment. The toughest little stone-thrower is in reality deeply vulnerable, not only to the Army or the IRA, but also to his own immaturity. He does not only need to be protected from terrorists and the security forces; he needs to be protected from himself.

The Media: How important is television as a behavioural model? A number of reports now available draw attention to the high violence content of both films and newsreels.[30-32] It has been fairly easy for researchers in this field to draw widespread public reaction, when this is their purpose, by simply compiling a list of the numbers of murders, robberies, muggings and gun-fights portrayed on the screen in any given year. By adding together a sufficient number of years and programmes it is possible to reach astronomical figures. The fallacy implicit in using these figures to shock is this: it presupposes, quite without evidence, that any effects are cumulative. These figures cannot fairly be coupled with the stock denunciation of broadcasting policy unless they are accompanied by experimental proof that the effects of TV violence are harmful and that they *are* cumulative. (They could, for example, be aversive, so *decreasing* the tendency to violence.)

At the same time, it is impossible to escape a feeling of nagging unease. Does research provide any answers?

In general, experimental work has shown that viewing a film depicting aggression *does* make children more likely to behave aggressively in the period immediately following; there is a high chance that, shortly afterwards, they will imitate the aggressor portrayed on film. Reviewing several such studies, Berkowitz concludes that they 'only demonstrate short-range effects. They do not show media-induced changes in *persistent* modes of conduct.'[4] Imitative aggression has, however, been shown to be most likely where the action of film is closest to the real-life situation. The child is more likely to behave aggressively, for example, towards a person who, in various ways, resembles the person attacked in the film. This tendency has been observed where the resemblance is by name or in appearance.[33] We come back, then, to an important point made earlier. Portrayal of aggression against a mythical bogeyman, such as a Red Indian or a rustler, is unlikely to have harmful effects on the child—perhaps the reverse. But a bogeyman in the community is a different matter. Taken together, all these results suggest, in fact, that by watching a riot on a newsreel the young viewer in an area where rioting is common becomes more likely to take part in street rioting soon afterwards. The BBC shows appreciation of this risk in its policy of not revealing the exact location of rioting shown on newsreels or reported on radio.

In his most famous allegory, Plato described a race of men who spent their lives, from birth, watching shadows thrown on the walls of their cave by a fire. Later, when they came out, they rejected the real world because it didn't correspond to the flickering shadows.

Too much attention given to the possible effects of TV violence after five years of *real* violence could attract, on the same principle, charges of shadow-boxing. Nearly an entire population of children now in the years approaching puberty can have no real memory of life outside a context of sporadic violence, barricades, fear and the constant tension and vigilance that goes along with it. In Chapters 5 and 6, I gave accounts of short-term symptoms of anxiety in most children, and medium-term symptoms in a very few vulnerable children. But has the long-term exposure to, and participation in, violence made any impact on these children? Has it, for example, made any difference in the way they behave at home and at school? What sort of adults will they become?

These questions are now posed more often and, when they are, another hoary old myth is re-told.

'Joe is eleven years old . . . He is a rioter—a mini-rioter . . . Social workers in the teeming, strife-torn streets of Belfast's "riot quarter" are concerned about the future of children like Joe. But . . . according to [a] social worker the prospects for Joe are not as bad as might be imagined. She says: "Pubescent boys are naturally rebellious and prone to delinquency in one form or another . . . I believe that when normality returns, these boys who have taken part in riots will have got such emotions out of their systems, and grow up naturally." ' (*Belfast Newsletter*, 30 October 1970)

If it has nothing else, the myth of catharsis has at least got respectable origins. Aristotle, in the fourth century BC, recorded his belief that tragedy on stage by 'arousing pity and fear' would result in 'purgation of such emotions'.[34] In other words, the act of witnessing or participating in a violent scene would make the subject less likely to behave aggressively afterwards. This concept, at best speculative and at worst dangerous, has been passed down to us in various forms and by various writers who have relied on little more than narrative ability. It is sometimes called the 'hydraulic' notion of aggression: the frustrated, angry, hostile adult or child, in an aggressive action, reduces his level of aggression by, presumably, a process of depletion. Konrad Lorenz, for example, tells of members of polar expeditions who, angered and frustrated by constant close contact with people they dislike, can steal out into the Arctic night, demolish an igloo with an ice-axe, then return to their tent and sleep soundly.

This tendency to 'feel better' after acting out feelings of aggression is, of course, universal. But does not this feeling in itself, acting as a reward, make aggressive behaviour more likely on a future, similar occasion? Research, in fact, strongly suggests that it does. Detailed reviews of work in this field are available elsewhere, and will not be reproduced in full here.[6][19,27,33,35] In short, the main hypotheses that are supported by controlled experiments are these:

Childhood aggressive games *heighten* the likelihood of attack on another child later.

The sight of a weapon, or possession of a toy gun, acting as an 'instigating stimulus', can trigger aggressive behaviour.

Watching a scene, in real life or on film, in which violence is portrayed, makes a child more likely to behave similarly soon afterwards. As suggested earlier, this likelihood increases where the later situation most closely resembles the one witnessed.

Aggressive play or behaviour has *no value* in reducing the drive to aggression.

Summing up a review of all available work on the question of catharsis, Berkowitz writes: ' . . . these negative results appear to question . . . the proposition that aggressive "energy" can be channelled off in a variety of basically equivalent ways . . .

'It is possible to maintain that aggression is a *learned habit* . . . [my italics]

'In general, there is no equivocal evidence of a cathartic lessening in the strength of aggressive tendencies following the performance of hostile acts.'[4]

I suggested, a page or two back, that a theory based on the idea of catharsis is dangerous—especially as applied to a situation of urban violence. It is dangerous, in the first place, because of the presupposition of a fixed amount of aggressive energy in a child's emotional make-up which can be drained out of his system as he finds outlets in sport or competitive games. This brings in its wake proposals of panaceas for riotous youth such as provision of more playing-fields or adventure playgrounds. Certainly, this need exists. And it is true that, deprived of a reasonable amount of playing space, a youngster may well express his anger by behaving aggressively; also, that lack of adequate recreational facilities will leave him with more time to roam the streets and get into trouble. But there can be little hope of success for any policy based on a notion that fear provoked by the presence of soldiers or anger following social or economic deprivation can be worked off on the football field.

The other danger consists in permitting children absolute freedom in expression of aggression in the belief that it will be drained away from a finite pool of this emotion leaving, in its wake, sober, responsible citizens. But, as experiments and experience in settings of urban violence suggest, an almost unlimited pool of psychic energy is available for translation into aggression, and this process is facilitated mainly by patterning on adult behaviour, by the presence of cues to aggression, and by adult approbation in the form of non-intervention.

More specifically, how have these longer-term *sequelae* of violence been evident in Belfast children?

During the past five years I have interviewed about 250 children who have been involved in rioting—50 at the Child Guidance Clinic and about 200 in the streets, in their homes, in youth clubs and youth organisations. Of the 150 that I have been able to follow during this period, there has not been one in whom aggression has shown a tendency to diminish with rioting. On the contrary, anger, aggressive talk and behaviour have increased; the children have progressed from stones and sticks to petrol-bombs and gelignite, and have often graduated from the street gangs to the Fianna or the IRA itself. A recent survey has reported similarly; boys who have taken part in one street riot are more likely to take part in riots subsequently.[6] A NSPCC inspector working in the ghetto areas confirms my own impression that stone-throwing at soldiers by gangs of children is *more* frequent, although now rarely reported in the papers; it just isn't news any more.

Also, since the rioting began, there has been a massive escalation of gang warfare among the 10 to 18 age group *within* the ghettos. There are now about 15 major gangs in North and West Belfast, each with 200 to 300 members; they are identified by different tartans (usually worn as scarves), and have a complex system of signs of recognition and cues to aggression. For example, giving a 'nod' to a member of another gang is an important signal of friendship; on the other hand, to wave the end of one's scarf is to provoke a degree of fury that has to be seen to be believed.

The important point here is that violence directed towards an outside object has been followed by violence *within* the group. This is only one facet of a general principle that violence, if habitually used to cope with one situation, tends to generalise to other settings and to other relationships. This is evident in all major sectors of these children's lives.

'This is a bomb hoax. You have five minutes to get out.'

Couched in a hoarse treble, the phone call came to a Belfast school recently—one of the three or four that he gets daily, the headmaster told me. And they are usually more convincing.

To a schoolboy who has, in his own eyes, successfully defied the armed might of the British Army, parents and teachers are small fry indeed. A naïve assumption, perhaps, if we are to go no deeper than surface probability. But a valid and age-old principle is involved, one that operates just as predictably in the community as in the laboratory. I call it the Macbeth Syndrome—although Macbeth discovered the underlying principle a little too late.[36] However, other leaders and community workers, as well as all shades of revolutionaries, tend to be equally disconcerted by the ease and speed with which anti-authority attitudes *generalise*—in due course to include the person who has taught them, or effectively condoned them by turning a blind eye. This is why, at the end of the day, groups who have arrogated power by violence have to use much harsher disciplinary measures than those used by the regimes they have supplanted—tarring and feathering, maiming, shooting. Aggression, as I have tried to show, is not an instinct and, as such, to be allowed free rein, but a *learnt behaviour* which becomes, when allowed, more and more the adult's or child's major means of dealing with all problems. From behaving violently towards the cultural bogeyman he turns aggression, finally, on to his own parents, teachers and friends—people *within* his own group, not identified with the enemy.

Teaching violence does not mean, in all cases, specific incitement. It does not always mean handing a child a gun, a petrol-bomb or a stick of gelignite. This is done, of course, but only by a very few adults indeed.

But what is the effect of violent words on TV, in the pulpit, or at home? What is the effect of stone-throwing in the street shrugged away, dismissed by old saws like 'They're only young once; boys will be boys'? A teacher once said to me, 'I can't stop them; that's not my job. Education's my job.'

The effect is, as I have suggested already, disastrous. We have seen, both in psychological experiments and in the community, the fundamental effect of cues from adults on childhood aggression. How are the responses to these cues carried over into a child's school life? The sorry tale of diminishing attendances, falling academic standards and increasing vandalism has now been told publicly many times by parents, headmasters, educational psychologists and educational welfare officers.[37-41] These confirm the accounts I have had myself from parents and children. Schoolwork has suffered very badly indeed and in 1972 exam results were the worst ever. This could hardly be otherwise, since Belfast children have now so many factors militating against satisfactory, or indeed any, school progress. Encounters with the Army on the way to and from school are usual; schoolbags are searched, and several youngsters I know have been summarily 'lifted' and have spent some hours in police stations being questioned—often by rather unconventional methods. (A soldier, having shown a young boy his gun and gained his interest, may ask casually, 'Have you seen one like this before? Has your daddy got one in the house?' On one occasion several children were left alone in a room which must have been bugged, since a police officer later referred, indiscreetly, to something the children had said when he had been in another room.) Apart from this, children have often lost sleep on the previous night because of riots, searches, or explosions and may be overtired the next day. Some have even fallen asleep in class and have had to be sent home. Population shift, resettlement and squatting have meant many children crowded into new schools, and some in no school at all.

Several headmasters in Belfast and Derry believe that the children are much more aggressive and less ready to accept discipline or to concentrate for long on any structured activity. A Derry headmaster said that even when they start school now, the children are 'old before their time', pugnacious and destructive—girls as well as boys.

'We hadn't meant to go on strike today, really,' a Belfast schoolboy admitted to me, 'but the girls from the Convent opposite came into the school yard and shouted to us to come out and join them. They shouted, "You're yellow, you're shit-scared," and overturned two of the masters' cars. We had to go down then. The masters locked all the doors so we broke the windows and got out that way.'

Attendance is down to 70% on an average in most schools in troubled areas, and sometimes to 50% or less. Vandalism has escalated frighteningly, with its 'multiplier' effect on environment. A visiting student spoke to me of her horror when, passing what she took to be a deserted ruin, she heard a bell ring and saw schoolchildren pouring out from their lessons.

To set against all this negative material—I was deeply impressed by what I heard from the headmaster of one Belfast primary school. In the heart of a disturbed area, the school building has twice been virtually wrecked by explosions. However, right at the beginning of the trouble, he and his staff decided to outlaw from the school all aggressive play and talk, toy guns, any references to or drawings of the riots. In sharp contrast with other schools in similar areas, attendance has remained around the 90% mark; children often arrive at 8 a.m. and have to be, in the headmaster's words, 'almost beaten away at the end of the day'. This is not very surprising. These youngsters have found a haven in the midst of an area of preoccupation with violence—something we would all like.

It is even more interesting that this headmaster has no problem with vandalism in the school. Again illustrating the potency, for a child, of cues to aggression or non-aggression.

A professor of psychology from California had found in most Belfast children, however, a significant preoccupation with death and destruction,[7] just as I have done. What impressed this investigator most was, however, the purely negative reaction to her standard question: 'What if there were no rules?' In total contrast, she says, California children reply invariably, 'That'd be great'! My own interpretation is that these Belfast children have for too long tasted the sweets of complete freedom and found them bitter. Because, in some areas of life in Ulster, there *are* no rules.

By now, habituation to violence is inevitable. It could hardly be otherwise, in a setting where a boy without a rubber bullet is like one without a penknife and where youngsters have attended school with sticks of gelignite strapped for concealment between shirt and skin. Violence has become part and parcel of life. On the morning after internment, as Catholic anger erupted, Protestant children, from an adjacent school-wall, sang, 'Where's your Daddy gone?' (version of a popular song), and, after the Derry shootings when thirteen demonstrators died, Junior Orangemen sang:

'We've got one, we've got two, we've got thirteen more than you . . .'

In general, childhood anxiety is now much *less*—and this is not an encouraging sign. A strange opinion to come from a child psychiatrist, perhaps. But anxiety can be protective; it is often the last defence. When throwing a stone, a petrol-bomb, or a stick of gelignite becomes merely a habit—an action performed by a youngster almost casually, without fear or anger—then the risk to that youngster rises steeply.

To summarise, each type has its anti-type, and each child culture has its bogeyman, an image introduced and reinforced both at home and at school. The bogeyman, usually a semi-fictional character—a rustler, a Red Indian —is a useful scapegoat on to whom a child can project his undesirable qualities, and in whom he can see them defeated with the traditional 'happy ending' of stories and films.

The twin cultures in Ulster have socialised their children into positions where the bogeyman is, for the Catholic child, the British soldier, for the Protestant child, the Catholic. This has happened because the opposite group member is, to the adult, the perceived frustrator. The child's fear, when strong enough, is accompanied by fantasies of riddance. The factors that favour his acting these out in the form of aggressive behaviour are proximity of the bogeyman, lack of restraints, modelling on adult behaviour.

Street aggression seems to have no cathartic value: it does not extinguish as a child 'gets it out of his system'. Instead, as a learnt behaviour, violence increases and in due course spills over into other areas of a child's life —home and school.

The effects of TV violence, guns and riot games depend mostly on the degree of closeness of the game or film to the real-life situation. Barricade-building at play or possession of toy guns are unlikely to have any harmful effect on, say, a London child, but could well predispose a Belfast ghetto child to real-life violence.

Mainly, however, it is behavioural cues from trusted adults, at home and at school, that will determine a child's actions on the street.

9

Education for Aggro

In a 'mixed' area of Belfast, if you are there soon enough, you can see the School Patrols—several dozen troops with armoured cars, riot gear, bullet-proof vests. Their job is to maintain physical separation between Catholic and Protestant children as they go home from school. Troop-carriers block side-streets, steel barriers are erected, then the children are released from the school gates and walked home in thin files, armed soldiers on each side. Protestants on one pavement, Catholics on the other. Why?

Sometimes the Army doesn't get there soon enough, and by the time the Patrol arrives bricks, stones and bottles are flying across the street. Some children have to be taken to hospital. Treble voices shrill, 'Fenian scum . . . Protestant bastards . . .' Again—why?

Now, virtually the only point at which Protestant and Catholic children can meet, or see one another, is across a street barricade or through the barbed wire of the Peace Line. The practical end-result of Ulster education-al policy is this: sectarian abuse is the only verbal exchange socially sanc-tioned, and rioting the only joint activity possible. This dismal Belfast street at four o'clock is our common campus.

At some stage, any writer on Ulster, or indeed on the problems of any other plural society, has to face, and avoid, the risk of finishing up with yet another well-I-never book of horror stories from the ghettos. He has to con-centrate instead on offering positive solutions. It is, in fact, in the firm belief that positive and practical solutions exist that I have written this book.

They are outside the scope of politics. There is nothing to be gained by trying, through legislation, to stop Protestants discriminating against Catholics, or vice versa. Prejudice is beyond the reach of laws and orders.

The search for fair government in Ulster's present, plural society is the search for a chimera. There can be no such thing.

Rash and sweeping words, perhaps. But in the last two chapters I have tried to depict the process of discrimination against a minority not primarily as something to be deplored or outlawed by condemnation—but something which is *inevitable*, given certain conditions. I have shown, I hope, that in any part of the world where an ethnic minority has been unable, for various reasons, to integrate with the majority, it in due course attracts discrimination, and in due course protests. The flash-point comes when both the size and economic situation of the minority begins to *approach* that of the majority. When a minority is very small, or very poor, the regime can afford to be generous. But as it grows, so does its threat to the *status quo*, and so does political and economic suppression by the majority.

My hypothesis is, in fact, this: the trouble in Ulster is not that the Protestants have been unfair to the Catholics, or the Catholics hostile to the Protestants. The trouble is that people think in these terms *at all*.

I am not, of course, suggesting that, to solve the Ulster problem, everyone should give up their religion. This is hardly necessary; after all, Anglicans and Methodists live in perfect harmony, in Ireland as elsewhere. But, as I emphasise from an earlier chapter, it is a *racial* problem we have to solve.

The role of segregated schooling in perpetuating racial divisions has been widely recognised. In this chapter I hope to trace the way in which Irish education has become almost totally segregated, to demonstrate its effects, and to put the case for primary school integration as the one initiative that would contribute more than any other single factor to the prospect of peace in Ulster.

First, to set against the anecdotes at the beginning of this chapter—a ray of hope from another plural society. In a San Francisco school, a programme of nursery school integration began in 1966, involving 60% Black children, 30% Whites and 10% Oriental children. A survey of attitudinal change among parents of these children included this, fairly typical, response from a White mother: 'When he came home and told me about Jack, Joe or Bill, I always asked, "Is he Black or White?" And, do you know, he had to *think* before he answered. After a while I got ashamed to ask. Now it's almost got that I don't care.'

An example of education in reverse? But then, children need to be taught bigotry. Could a result like this be achieved in Ulster?

Separate education for Catholics is nothing new in Ireland—or anywhere else. The Catholic Church has concentrated on providing schools ever since universal education began—although the Church's proscription of attendance at non-Catholic schools was only clearly formulated, at least for the British Isles, in the nineteenth century, following a failure of the 1807 Parochial Schools Bill to bring in a system of universal non-denominational education. Later, in 1929, Pius XI condemned 'neutral or secular' schools (as well as co-educational schools and sex education). The Church's present position is defined in the 'Declaration on Christian Education' of Vatican II (norm *ix*): 'Catholic parents [have a] duty to entrust their children to Catholic schools, when and where this is possible.' The pre-Conciliar Canon Law of the Church proscribed attendance at '. . . non-Catholic, neutral or mixed schools, that is schools that are open also to non-Catholics . . .' (Canon 1374). But it has been pointed out that this proscription is not absolute; it applies where the schools concerned are a 'danger to faith or morals'.[1]

In Ireland the Catholic Church has always maintained its own schools and, even before partition, there was, in 1831-5 and 1919-23, fierce opposition on the part of Catholic bishops to attempts to set up an integrated State education system. The Belfast Education Bill of 1919, for example, one such proposal, drew immediate and total condemnation from the Northern hierarchy, who saw it as an attempt to influence children in favour of partition and Unionism. The British Government's Education Bill of 1920 suffered the same fate: Irish nationalists feared, perhaps not without reason, domination of education by Orange influence.

After partition, Ulster's new Ministry of Education produced a remarkable and far-sighted document, the Education Act (N.I.), which received Royal assent in 1923. Among its provisions, all schools were to be transferred to local education committees, religious education was forbidden within school hours, and the education authority was not entitled to take an applicant's religion into account when making an appointment. The Minister of Education, Lord Londonderry, told Church representatives that 'all the quarrels between Roman Catholics and Protestants arose out of the teaching of the Bible and as he wished the children of all denominations to meet in the same schools and grow up in a friendly atmosphere he thought this could only be achieved if there were no Bible instruction and if Roman Catholic and Protestant children mixed in the same schools.'[2]

All the Ulster Churches opposed the Act; the Government wavered, and the day was lost. Intense pressure from a united committee of the Protestant Churches and the Orange Order succeeded, in amending Acts of 1925 and 1930, in removing these two secularising sections from the 1923 Act and in securing the representation of Protestant clergy on school committees. The

Catholic Church, in part continuing earlier policies but also probably fore-
seeing the success of Orange pressure to amendment, refused from the beginning to transfer any schools to local education committees.
Since then, both Churches have stood firm in retaining rigid control of
the two parallel education systems, Catholic and State. 98% of Ulster children are at segregated schools. Financially, the Catholic schools (about 680
in all) receive 65% State grants for building and maintenance, rising to 80%
where the school accepts 'maintained' status under an Act of 1968 (i.e. one-
third public representation on its management committee). At present, 524
Catholic schools are 'maintained'.
    All parties remain dissatisfied. The Orange Order and Protestant Church
leaders complain bitterly about what they see as undue generosity on the
part of the Government to a system over which it has little or no control.[3,4]
In return, the Catholic Church points out that its members pay the same
amounts in taxes and local rates as Protestants but receive only partial sub-
sidies while, at the same time, subsidising fully a State system from which
they have chosen to secede.[5]

    It is important, before the arguments on education are developed any
further, to appreciate the special role of the school in the clusters of village
communities that are Belfast and Londonderry. The church, with its nearby
school, is at the centre of the community, spiritually and geographically.
Pupils from the ghetto area that the school serves travel the short distance
from home to school, form their friendships there, learn there their games
and develop their interests, play in the nearby playgrounds and, later, look
to their clergy, parents and teachers to place them with sympathetic emplo-
yers. If they themselves want to teach they will go to segregated colleges and
then return to the ghetto school, completing the full circle. Even in Belfast
University, a liberal enough campus where Protestants and Catholics are
supposed to mix equally, polarisation is in fact the rule. Affinity groups tend
to be almost totally segregated by religion—suggesting that paths have
diverged long before.
    It is not very surprising that integrated housing and employment schemes
run into difficulties when, as early as adolescence, the consciousness of two
cultures is so deeply interwoven as to produce, already, this degree of
mutual alienation. What has happened?

    A positive programme of socialisation into an ethnic group is not neces-
sary to produce at least some degree of consciousness of kind and, at a later
stage, aggressiveness towards other groups. This has been demonstrated in

several cross-cultural studies; one is described on p. 117. Another group of investigators, studying the behaviour of Black and White schoolboys at a summer camp, emphasised (and were clearly disconcerted by) the tension that could develop between segregated groups—tension that could flare into anger if an outside source of frustration was introduced.

Church leaders in Ulster dislike the term 'segregation'; they assert that the system is simply one of 'separate education'. Children, we are told, can play in the streets after school, since no prohibition is placed on this. 'Unfortunate nonsense,' wrote a teacher in the *Irish News*.[6] Unfortunate because it sounds like the sweetest of reasons to any reader who does not know Ulster. Nonsense because, in the ghetto system, it is a physical impossibility.

How far does indoctrination, political and racial, play a part in the education of Belfast children? The word, unfortunately, is highly emotive, with its whiff of brain-washing, thought-control, and the other squalid techniques that rob an adult or child of a measure of his status as a human being. But, for the present purposes, to conclude that political indoctrination exists is not at the same time to argue conscious motive on the part of parents, teachers or clergy. This may exist. But in the main, pupil and teacher alike are involved in twin systems that have developed opposite momenta—where words have rationalised and glorified past actions and in turn provoked a further set of actions; systems where the essential corrective of contact with the feared group has been lacking—the only proven means to scotch myths and avoid stereotypes.

The misconceptions and prejudices held by children, as might be expected, mirror those of their elders—their only source of information in default of experience. To the Protestant child, the Catholic is work-shy, dishonest, dirty, a breeder of big families, productive of community poverty. Examples from my own files have already been quoted, and there are many other examples in print.[7]

In the Catholic child's attitude there is more conscious fear. The Protestant to him is the person who wants to burn him, shoot him, deprive him of his home or parents—the brutal oppressor. Even if one is familiar with Ulster, the expression of this myth can still come as a nasty shock. Very recently, with three young Catholic boys in the car, I stopped to watch a Loyalist rally. The youngsters, terrified, actually crawled as far as possible under the car seats to hide, and refused to emerge until I had moved on. Psychological studies now in progress may shortly tell us a good deal about how the child communities in Ulster perceive one another, but that one episode alone told me more than a year's work with questionnaires and punch-cards might have done. These three boys were approaching adolescence, but fear and hatred begin much earlier. Nursery-school teachers and parents have often expressed to me, as to others, their dismay at hearing four- and five-year-old children lisp lurid sectarian epithets. The same degree of unhappiness was conveyed in the report from Watts (p. 88), where stereotyped notions that Negroes were bad and violent were found to be prominent in children of four and five. Several other studies in situations of racial conflict have demonstrated the early acquisition of attitudes and stereotypes and the great difficulty of reversing them later.[8] It seems that a cut-off point exists, after which age neither explanation nor real-life contact with the feared group can modify the basic hostile stereotype; it seems also that it comes fairly early.[9,10] In Ulster, any experiments that have been possible in integrated settings (reviewed later) suggest that this stage is reached shortly before the age of puberty.

While motives may be open to question, the fact of socialisation into two different cultures, British and Irish, remains. This is evident even outside the school gates; Protestant youngsters are bedecked with Orange, Ulster and Vanguard badges and tartan scarves, Catholic children with green scarves, shamrock badges and Connolly badges (the latter *behind* the lapel).

As for games—was the battle of the Bogside won on the playing-fields of the Gaelic Athletic Association? A near-religious faith in the ability of organised games to mould national sentiment is as much part of Irish as of English folk-lore. So Sean plays hurley and Gaelic football, Billy plays soccer and rugby. (The hurley-stick, a familiar weapon in street riots, has to recommend it not only a useful solidity and leverage, but also a strong Gaelic cultural mystique.)

In the classroom, history lessons have always been based on two widely divergent curricula and sets of interpretations. A 1970 report reads: 'Protestant children [have] acquired an entirely different version of Irish history [from Catholic children], an emphasis on their plantation among a hostile and backward people, on the need for eternal vigilance, and on the siege of Derry . . .'[11]

An Irish branch of the European Association of Teachers gave its opinion, in 1967, that history teaching in Catholic schools was 'insular and essentially political'.

Barrett and Carter wrote, in 1962: ' . . . the non-Catholic schools teach English history as being a well-established discipline with good textbooks, which tell children about their own country [which is the United Kingdom] . . . Catholic schools are more likely to teach Irish history in its own right and to treat it as the story of heroism in maintaining national feeling under foreign rule.'[12]

The cultural scene expands to include other subjects and activities. In Catholic schools great importance is laid on the Irish language as a subject; children are taught Irish music and dancing and compete in the *feisanna* that abound locally. The strong national content of drama has already been illustrated (p. 118). In the Protestant schools, the youngsters are directed into the Orange Lodges and learn, mainly from their peers, the 'right' slogans and songs. Out of over a hundred Protestant schoolchildren whom I have spoken to, all have been able to sing, word for word, for example, the song quoted on p. 40 and all but a few know this, to the tune of a popular song:

> I'd like to buy the Pope a rope,
> And hang him from a tree,
> With Bernadette and Gerry Fitt,
> To keep him company.

All teachers, of course, claim to deplore this open political aggression. But a good deal of confusion of motive is apparent. The headmaster of one large Belfast Catholic school stressed the great importance and benefit of a 'Catholic atmosphere' (which it is not for me to dispute). However, the four school houses were named: McCracken, Casement, Hope, and Pearse[13]: three of these were Protestants, one a Presbyterian minister. Their common quality was, as it happens, fanatical anti-British activity. (This school was later used for an IRA Press conference.)

In another Catholic school, an Army search turned up collages, part of a set project displayed on the wall, made up of press photographs and headlines: the word 'murderer' across a soldier's back, 'kill' across a group of B-Specials, 'policeman shot dead in Strabane' juxtaposed with 'laugh'. It is also of interest that on the Queen's birthday in 1972 (a school holiday) several Belfast Catholic schools remained open.

This strange dualism mirrors almost exactly the Junior Lodges' cloaking of politics with stated religious aims; a motto, 'Civil and Religious Liberty', together with much display of clerical collars and open Bibles seems to accompany, without conscious unease, the most vicious of anti-Republican songs and slogans.

Some effects of political indoctrination have been shown in a study by an educationalist, Alan Robinson, in *WHERE* (published by the Cambridge Advisory Centre for Education). For example, of one thousand Derry children, the majority of Catholics named Dublin as the capital of their country; all the Protestants named Belfast. Again, asked who the leading citizen of Derry was, the majority of Protestants named the Unionist mayor. Most Catholic children named the Bishop or the MP (a non-Unionist). Robinson adds: '. . . the fundamental division of seven-year-olds may be expected. But the fact that this dichotomy is even more marked at 15 years is alarming.'[14]

Robinson blamed all of this information and misinformation on teacher bias and in doing so, I think, overstated his case. But there is little doubt that teachers have, even if unintentionally, succeeded in conveying some political doctrine to their pupils. In Rose's study, 62% of Protestant adults and 58% of Catholic adults were able to recall their teachers' political viewpoints.[15]

A study which has just become available, carried out for the Community Relations Commission, found similar differences of national and political identity. Three-quarters of Ulster Catholic boys feel they are Irish; Protestant boys are equally split between being Ulstermen and British. In a question about who the children thought were 'people like us', most Protestant children identified with England, and most Catholic children with Eire.

This investigator also found that the children who identified most strongly with one nationality or the other were those most likely to approve of violence for political ends. Also, *the children without friends of the other religion were much more likely to endorse political violence.* In the case of Protestants, this difference between those who had, and those who had not, got Catholic friends was 'overwhelming'.[16]

If the present, segregated, school system is a potent factor in perpetuating community differences, the implied corollary is that integration would prevent or modify these differences and the consequent aggression. Is it as simple as this? It has been claimed, for example, that as these children's attitudes mirror their parents', it is necessary to change *their* attitudes first, perhaps by integrated planning in housing or employment. If this were feasible, of course, all would be well, but attempts in this direction have failed signally and disastrously. (Unity Flats, New Barnsley Estate and even Ballymurphy started life as joint Catholic-Protestant housing ventures.) The argument is sterile anyway, since the parents themselves were at separate schools and are thus products of segregated systems; it is a chicken-and-egg dispute where nobody wins. Unfortunately, several attitude surveys, in so far as they were attempts to assess the relative importance of home and school influences in fostering prejudice, have seemed bent on the curious exercise of tracing a 'vicious circle' back to its source. It would be more worthwhile to ask whether, and where, the circle might be broken.

Unfortunately, too, whatever prejudices parents may convey, the neighbourhood school does nothing to eradicate them; rather the attitudes are hardened and the myths rendered more florid from being shared among classmates. (These are probably the child's major learning source in school. It is a pity that all attitude surveys have ignored this potent influence, concentrating instead on teachers. Most parents, genuinely I think, express themselves horrified by the aggressions, the words, and the songs that their children learn from schoolfriends.)

But the fact remains that, whatever the school's role as indoctrinator, it has, primarily, been the parting of the ways for adult and child alike —physically, if nothing more, the *fons et origo* of the ghetto community.

What could school integration achieve now? The results of many years of experimentation in other racial contexts are available to us. Berkowitz, for example, reviews all the strategies that have been used to prevent or reverse formation of ethnic prejudice and hatred.[17] He asks, 'How can men of goodwill proceed to lessen tension between groups?' The most difficult problem is to prevent frustration-induced anger from bursting into aggression against more or less innocent victims. However, 'hostility against people stemming solely from their group membership can perhaps be lessened fairly easily'.

The methods he reviews are, first, 'communications advocating peace and harmony'. From an extensive literature on the use of communication techniques for the lessening of inter-group hostility, the conclusion is that exhortation is useless; worse than useless, in fact, since the peacemaker is invariably seen by each group as supporting the other side ('nigger-lover', 'Fenian-lover', 'Uncle Tom' are just about the most scathing epithets in the English language). Further, scaremongering or indicating the dire effects of continued aggression are just as futile; aggression *increases* as the level of fear and anxiety is raised. External threats are just as likely to polarise as to break down barriers.

Berkowitz concludes that, across racial divides, *equal status contact* is the only potent means of reducing hostility, particularly between children. One study, among 106 White boys at an inter-racial summer camp, demonstrated increased friendliness towards Negroes after the camp. 'Equal status contacts between groups . . . lower inter-group enmity, particularly if informal social relationships are involved.'

This assertion has been amply proved both in school and in the wider community. A project for nursery school desegregation in the United States, supported by the National Institute for Mental Health, found that, in a group of sixty children who had attended the racially-mixed schools from the age of two, all were *without* racial prejudice at the age of five—in sharp contrast to the consistent finding of racial stereotype and hatred among four- and five-year-olds in segregated schools.[18] Where parents were involved, this process of education could even work in reverse (p. 135).

Professor Alfred McClung Lee, one of America's leading authorities on race relations, writes: 'Through the casual experience of classrooms and playgrounds, White and Negro children can learn to associate with one another ... without antagonistic racial frictions ... "Going to school together" works powerfully against intolerance.'[19] In London's Notting Hill, too, a ten-year programme of school integration has largely eliminated aggression against minority groups—including the Irish.

Opportunities for studying effects of integration among Ulster children have been regrettably limited, but what evidence we have is encouraging. For example, seven of Belfast's eight special schools (for handicapped, educationally subnormal and delicate children) provide for both Catholics and Protestants. The headmistress of one told me that there was no evidence of polarisation or aggression among the children, who have mostly been there since the age of five. She feels that they are generally less prejudiced than most children, both inside and outside school. In another integrated school, for maladjusted children of generally average intelligence, there has been quite a lot more in the way of name-calling and sectarian fighting, particularly in the ten-plus group. But this is only evident in newcomers; within a few weeks firm friendships are formed across ethnic boundaries—a microcosm of what *could*, hopefully, happen in the general school population.

But as things stand, it seems to me deeply ironic that the privilege of mixing with children of other religions is available to these handicapped groups alone. I am left with the shattering and brutal conclusion that these children must be thought too stupid to be corrupted by integration.

Within the last two years, various projects involving joint activities for Belfast schoolchildren, such as those run by student volunteer groups and by the NSPCC, have largely broken down. But before this, and from my own six-year-long involvement in youth organisations, it was clear that the same age differences existed as in the special schools. The prejudices of younger children are soon much less in evidence, and friendships formed at camp or in other joint activities often continue for long afterwards. In the thirteen-plus group, however, polarisation and aggression is much more evident; at this stage there seems to be no strategy that can counter a too-long exposure to more powerful influences. The best that can be hoped for, I have found, is a guarded tolerance, but that can quickly break down when things go wrong. Any undesirable qualities seen in the youngster's opposite number are then put down, automatically, to his group membership. A boy thought dishonest is a 'Fenian cheat' or a 'dirty Prod'. On the other hand, one who has likeable qualities is, by this age-group, seen as a rare exception. 'I don't think of Sean as a Catholic' is a characteristic, half-surprised comment. Things will go more easily for Sean, perhaps, but his friend's concept of his member-group as a whole remains untouched. Further, any friendships between members of this older age-group do not, in general, persist beyond the period of the joint activity.

But the younger child is more susceptible to modification of his stereotype. Not until this period of first-hand experience has he had the opportunity of making up his own mind about the 'others'. His view has been flat and one-sided, a cardboard ogre made up of undesirable qualities only. Now he can learn that there are also likeable features—more than sufficient 'sameness' qualities to balance the alien qualities. Even now, I still cannot quite get accustomed to the sense of surprise communicated by, say, a Protestant child who discovers how his counterpart likes the same food and drinks, enjoys, like him, playing in trees and water and getting in a mess, and that he has the same rows with his parents over the same things.

But the child nearing puberty has by then, it seems, formed perceptual barriers that are virtually impenetrable.

As a strategy for reduction of ethnic prejudice among children, this casual-contact method may sound idealistic, even ingenuous, but it has been found to work in practice. Circumstances have limited the amount of research possible, but there is now sufficient evidence to pilot, without misgivings, a wide programme of primary school integration.

Practical difficulties apart, arguments by powerful voices have been heard against integration, and in favour of continued separate education for Catholic and Protestant schoolchildren. As always in Ireland, the Church leaders have been the main contenders. The Protestant–Orange group has from the beginning eliminated the possibility (remote anyhow) of Catholic cooperation by its insistence that children in State schools must be taught the 'Protestant faith' in school hours, by maintenance of a minimum of 50% representation on the Education Committees by Protestant Churches, and by unwillingness to employ Catholics at all levels.

The Catholic Church has been even more vocal. A recent booklet by Cardinal William Conway, Primate of All Ireland must, one presumes, be taken as the Church's official view on the subject. As such, it deserves some attention.[20]

The first of his arguments is that, among advocates of integration, motives are highly suspect. He suggests that they are trying to blame all of the present unrest on the Catholic Church by way of an attack on its educational policy. He quotes from a visiting journalist: 'At the moment the only people who are professing wholesale desegregation are extreme Unionists who believe that this is as good a stick as any to beat the Catholics with.'

It is a pity this anonymous writer did not do his homework more carefully; if he had he would not have beguiled the Cardinal into such a misconception. The advocates of integration, in fact include the following: The National Council of Civil Liberties and the Northern Ireland Civil Rights Association;[21] The Cameron Report;[22] The Catholic Renewal Movement;[37] The Belfast Humanist Group;[23] Official Sinn Fein; The Alliance Party;[24] The Northern Ireland Labour Party;[25] major reports and surveys in *The Times*,[26] the *Economist*,[27] the *Belfast Telegraph*,[28] *WHERE*;[14] two-thirds of the total electorate, both Catholic and Protestant;[29] pupils in Belfast grammar schools as shown by two independent surveys;[30,31] the Irish Communist Party;[32] Miss Bernadette Devlin;[33] Captain Terence O'Neill (now Lord O'Neill of Maine),[34] The Ulster Teachers' Union, The The Irish National Teachers' Organisation.[35]

Frankly, I have never heard an extreme Protestant advocate desegregation—and I doubt whether the Cardinal is in the confidence of this group.

The Cardinal's second argument concerns the importance of the 'religious atmosphere' in a school. It is argued, without evidence, that home influences are not sufficient to ensure that the child grows up a devout Catholic.

But several studies, both in England and in the USA, have shown that the major factor in determining later observance is the importance placed on it *at home*, rather than at school. Religious observance in childhood and later is no different between those who have attended Catholic schools and those who have not.[36-38] It could also be argued, with justice, that segregation actually represents a *failure* of moral education, rather than the reverse; certainly, advocacy of the policy goes strangely with this paragraph: 'We must love our fellow-men, not because of some kind of herd-instinct, but because "every man is my brother", a child of the same father who loves us both.'

The Cardinal argues finally that separate schools for Catholics and Protestants have not produced separate communities in other countries, such as Holland or the USA. This, of course, is true, but segregation there is by religion only; children at both Catholic and Protestant schools, in these countries, share the same race and allegiance. They have more in common than they have differences. But where segregation has been racial/political as in Ulster, it has in the short term perpetuated the fiercest of aggression between young children, and, in the long term, been the origin of the ghetto, and of the 'mutual fear' held by Scarman to have been the main cause of the disturbances from 1969 onwards.[39]

Of all major academic writers on the Ulster situation, Professor Richard Rose is the only one, to my knowledge, who has been doubtful about the possible benefits of integrated schooling.[15]

In the first place, he says, 'conflict in Ulster antedates by two centuries or more the introduction of separate and compulsory education'. It is correct that separate education as a stated *policy* is comparatively recent, but *de facto* segregation dates from the seventeenth-century Plantation of Ulster, when the communities and educational establishments of the settlers remained both culturally and geographically distinct from the Irish settlements and the Church schools; the opportunity to integrate was never available.

Rose says, further: '. . . a better understanding of the opposite religion will not necessarily lead to greater trust. A Catholic at a mixed school may learn that when Protestants say "not an inch" they mean it, just as a Protestant may learn that his Catholic classmates refuse to regard the Union Jack as the flag to which they give allegiance.'

This is one of the few pieces of speculation in a study otherwise carefully pegged down to objective fact. It is also based, unhappily, on a common logical flaw. The writer has visualised each child purely in terms of his stereotype—as the cardboard figure composed of all the features *unattractive* to the other group. But these do not, of course, fully describe a child. As we have seen, in an integrated setting the *common* features in practice far outweigh those that are alien; the discovery of shared characteristics can quickly relegate symbols and slogans to their proper place.

Rose has also, I feel, been too restrictive in limiting his discussion to the role of learning in the classroom alone, and in not giving sufficient attention to the school as a growing point for the community, the place where children will find their lifelong identity and affinities.

Segregation exists in more than one form. For example, the social barrier immediately below the middle class is probably higher in Ulster than in any other Western community. An almost total lack of comprehensive education leaves the grammar-schools as overwhelmingly middle-class preserves, and a serious inequality of opportunity has resulted—in an indirect way. It is true that any child with the necessary academic ability can obtain a free place at a grammar-school but, at the same time, the clash of social standards between school and home has led, in my experience, to numerous psychiatric casualties. Not only does the physical milieu of the ghetto home make academic progress difficult, but the child can become emotionally isolated both from his family and his middle-class schoolfellows. Some can cross the barrier fairly easily, others live a sort of double life, but a great many whom I have seen have built up a huge load of resentment to classmates and to school staff who, they believe, feel called upon to impart middle-class standards of dress, talk and behaviour as much as, or even more than, academic knowledge. These children, in adolescence, can develop an almost pathological aversion to uniform, prefects, compulsory games, and the other traditional features of the grammar-school. Generally their attainments lag far behind their ability, and most of this group leave school early. This is in line with the conclusions of Jackson and Marsden, who found, in the North of England, that wastage from grammar-schools was largely a *class* wastage.[40] The difficulty seems even more pronounced in Ulster, where there are still fewer opportunities for upward social mobility. Much more has to be done in the way of providing a full academic curriculum for a child *where he is,* socially as well as geographically.

To segregation by religion and social class most Ulster schools have added segregation by sex. Few psychiatrists would hold a brief either way on the latter; nevertheless, all these divisions make for fragmentation of resources as well as lack of opportunity for social enrichment. Where children are educated in these 'little boxes', artificial communities form and own-group overvaluation results. In Belfast, the religious division means a loss of opportunity for many Catholic children, for whom grammar-school places are in short supply. At the same time, several State grammar-schools have a regular annual quota of unfilled places. The double bar to entry for the able Catholic ghetto child results in these places being taken by less able children, unqualified 'paying' pupils.

Integration of Protestant and Catholic children will not come overnight; indeed, it may never come if the decision remains in the hands of Church leaders with an interest in maintaining numbers in the pew. I have tried to point out that other interests exist, and that they do *not* necessarily run counter to those of the churchmen, as they believe.

Pressure to integrate could, I believe, come most effectively from parents. Many Catholic parents to whom I have spoken would, for example, like to send their child to the nearby State school, but hesitate to make him a human sacrifice. They wait until several friends may be able to go with him, and the opportunity passes. Certainly, if school integration were accepted in principle, elaborate busing programmes would not be required, since physical distances between ghettos, in contrast to the USA, are generally small.

Failing total integration, the common campus remains a possibility. Here, while the children had lessons in separate classrooms, they could meet for games and extra-mural activities. But I am bound to say that the common campus only makes sense as an intermediate stage rather than as an alternative. It would spotlight a philosophy whose absurdity could no longer be ignored.

As things stand, I risk the following predictions. First, that there can be no purely political solution to the Ulster problem. Second, that total integration of children from primary-school-age upwards would be the most potent single factor in breaking down community barriers and in restoring long-term peace. Finally—always provided that the human animal remains unchanged—while segregation of schoolchildren continues, episodes of community strife in Ireland will recur throughout, and probably beyond, the foreseeable future.

10

Prospect

Even for a very sick society, powerful medicines may exist. In the previous chapter, for example, I suggested that a programme of non-denominational and comprehensive education in Ulster would be a strong weapon to break down cultural and religious prejudice. After this—and only after—Ireland united under a secular government, and perhaps with EEC investment in industry, would have an excellent chance of economic viability.

But this is a long way off. Unfortunately too, measures that seem less than radical tend to be scorned as palliatives, as appeasement, as sops to extremism. They may be. But no doctor despises palliatives. They can cushion a patient against the worst effects of his illness, they can halt a decline, or they can prepare him for surgery. With the community—in particular the child community—as patient, what measures are available to minimise the effects of violence within the present political framework?

All of Ulster's city children are at risk, but not equally. There is now enough information to enable us to sketch out, fairly consistently, the profiles of two children especially at risk.

The first, slightly more likely to be a girl than a boy, is aged about ten. She is probably a member of a large family with a low income, and has one parent who is absent from home or chronically ill. She lives in a 'riot area'; at least one of the adult figures in her immediate family is suffering from crippling symptoms of anxiety as a result. This adult is preoccupied with the troubles almost to the extent of being unable to talk or think about anything else; he or she is given to thinking aloud about the imminence of catastrophe, and is particularly prone to conveying helplessness and uncertainty: 'What will become of us . . .? They'll kill us all yet . . .'

The child herself has had more than the average number of physical illnesses. She has always been prone to over-anxiety and, early in the outbreak of street trouble, shows a consistently heightened response to sights and sounds linked with the disorders—for example, to loud noises, or to crowds.

This link may be less direct. The young son of a member of the Ulster Defence Regiment fainted, then developed a choking fit when a car braked outside the front door and on another evening when the door was hammered by someone in the street. But there was no need for 'interpretation' to this boy by a therapist who had also read the horrifying Press reports of masked IRA assassination gangs.

The second youngster at risk is a boy in his early teens. He too is economically deprived. One parent is deeply involved in the militant Republican movement and at least one male relative is, or has been, interned. The boy is mentally absorbed by the conflict and has no other interests. He expresses extreme hostility not only to the security forces but also to other authority-figures, including his parents. He is the leader of a gang but has probably, so far, been quick-witted and nimble enough to avoid being 'picked up'.

What can be done for these children? There is a strong case for removing both from the area, at least for a short time. But this is an option to be used only with the greatest of discretion, particularly in the case of the child with emotional symptoms. It is worth repeating a point made in Chapter 6, that the *majority* of children will suffer greatly from being separated from their immediate family at a time of stress. Teachers in refugee centres have noticed, in particular, deep anxiety, bed-wetting, diminishing powers of concentration, and an almost pathological suspiciousness developing in children from the North. Also, more strikingly, several children developed squints. It is well recognised that a child's existing squint can become more pronounced with fatigue or depression, but this is the first report, to my knowledge, of squints appearing *de novo* in a group of children.

In short, removal of a child from his family in the riot area, barring physical danger, is only justified when the relation between the parent and the child is pathological. That is, when the parent is communicating to the child more anxiety than the latter can cope with. This is a very difficult assessment to make. Granted that adequate shelter resources were available (which they are not) selection of children for them would certainly be a job for a professional multi-disciplinary team. Failing this, the advice one can offer on evacuation of children is the simple 'If in doubt—don't.' The chances are that they will suffer for it.

Granted that the emotionally disturbed child can be admitted to a treatment centre—what then? One is often asked this: if the child's main fears are of damage to his home, injury to his parents or internment of his father, don't these risks remain the same even though he is away from home? Might his anxiety not increase? Also, he will have to go back to the area sometime; could it not be even harder then for him to adapt?

On the first question, while the risks certainly remain the same, the child's symptoms always, we have so far found, diminish rapidly in the altered environment. This alone argues strongly, again, that the reality factor has had little bearing on the child's anxiety; *communicated effect* has been its main determinant. It is not the event that has done the damage, but the parents' reaction to it. As for going back to the area, certainly it may not have changed, but one usually finds that the child has. Away from an atmosphere of preoccupation with fear and violence, he has found new perspectives, discovered new viewpoints, and has learned defences for coping with his anxieties. Support and advice on handling for the parents also help—but the important factor is the immense benefit the child *himself* receives from temporarily changed surroundings.

Apart from this special group, I feel—and several headmasters agree —that all youngsters will profit greatly by a short spell away from a troubled area—at a period when there is *not* serious rioting. Large groups of children are taken to the country by one headmaster who stresses the 'humanising effect of Nature'. Another puts the benefit of these outings down to a child's learning of 'rural values' which he feels are superior to urban ones. At a meeting where the subject was discussed, these flights into the metaphysical left me temporarily defeated, but I think that the advantages are real enough. The gains for the youngster are probably in the sense of proportion. The noisy street becomes at once only a small part of a larger, much more exciting world; there are suddenly other possibilities besides the regular, nightly game of stoning the foot patrol to a retreat behind massed Saracens.

The violent youngster has a crucial role in the street as the nucleus of his gang. The Army have always believed that their most potent strategy for crowd control consists in isolating these leaders. The 'snatch squads' operate on this principle. Russell has found, as I have, that these youngsters can be identified (apart from a riot situation) by the characteristics listed at the beginning of the chapter.[1] It makes sense that this youngster, too, should have priority in any holiday or evacuation scheme. He is the nucleus around which the gang forms. Two or three gangs are a mob, and two mobs are a riot.

This boy's removal is not suggested simply for the purpose of protecting property and other youngsters; he himself can be helped. But he presents a much more thorny problem than the child with anxiety symptoms—and for all sorts of reasons. The implications of offering 'treatment', for example, are often taken seriously amiss by members of a community believing themselves to be engaged in a legitimate war with the British troops. There is a belief that he is fighting for his country, for Ireland. By appearing to brand the youngster as psychiatrically ill, is the therapist taking sides?

He is not. He is concerned not with the child's politics or his nationality, but with his immaturity. For the reasons enumerated in Chapter 8, the child is as yet unable to appreciate the risks he is taking, and requires protection until he does. As it turns out, judicious management of this type of boy has much the same positive effect as it has in the anxious children—at least in those of whom I have had personal knowledge. It has proved possible to penetrate the child's obsession with violence and to introduce new interests and pursuits.

But this takes time. A group of student volunteers reviewed their summer's work in Ballymurphy with something like despair. In fact, I believe it *was* despair. On most Saturdays they organised a day at a seaside resort, with games and a picnic, for a hundred or so children, but when it came near 6 p.m. their charges would ask anxiously: 'Please, will we be back in time for the rioting?' And, for me, the tired and bewildered faces of the young students took any possible humour out of the story.

Another youth leader in the Falls area, having tried every conceivable form of activity, told me, 'There's no attraction like a riot,' and closed down. Was he right?

It is much too easy to condemn the crowds that gather where trouble breaks out. Few adults, let alone children, are possessed of enough restraint or detachment to avoid being where the action is, if only as spectators. The first hurdle of a youth leader working in such an area is acceptance of his limitations. Short of locking children up, it is probably impossible to keep them away from the scene of a riot, once it has started. But very many major disturbances are sparked off by groups of children throwing stones or paint at Army vehicles. It is obvious to any observer that, although it is not the whole story, many of these children engage in this kind of activity quite simply because they have nothing else to do. They roam the streets in huge, aimless gangs, the troop-carrier appears, stones are picked up . . .

So by providing other interests in the youngsters' home area a youth leader can avoid, at least, the moment when the Army comes as God's gift for an idle afternoon. While he can't hope to alter social and political outlooks (and would be exceedingly foolish to try) he has an immensely important task within these limits. In every side-street one sees boys in their middle or late teens simply bouncing a ball off a gable wall, hour after hour. For them it must seem, as things stand, a not unwelcome diversion to bounce a brick off an armoured car.

Teaching new means of enjoyment, altering the behaviour patterns of four years, is not the work of one afternoon. It takes not only time and patience, but a great deal of tolerance and flexibility on the part of the leader. The traditional youth organisations, in their present form, have for this reason little to offer ghetto children.

If a youth centre is to be built, the best way to insure against its not being almost immediately burnt down again is to let the youngsters build it themselves, supplying the material and instructions. In an underprivileged area near Dublin, a youth club that was built in this way several years ago is still guarded with fierce possessiveness and is the only building undamaged by vandals for miles around.

Short-term 'integration' projects involving both Catholic and Protestant children are of doubtful value. As studies to pilot integration of the wider child community these outings, camps, and 'twinning' of schools are of some use, but not as an end in themselves. It is true that by equal-status contacts children, especially younger children, can learn tolerance and unlearn myths. But they find their new tolerance little appreciated when they return home to communities where myths are the gospel and sectarian hatred the first of the commandments. In my experience, several of these children have become to some extent isolated from their peer groups, and a new clash of values has been created in their minds. Some, to their great credit, retain cross-religious friendships at considerable cost, in terms of victimisation, to themselves. Others find that they can only regain status by at least appearing to revert to the old songs and slogans.

Is this unfair to the various bodies that organise these projects? Perhaps —but if it is thought desirable that Catholic and Protestant children should mix, then the schools should be desegregated. If not, then resources should be directed to helping the casualties of the divided community—the children at risk.

Other irritants could be removed. For example, the Army's patrolling of ghetto areas seems a singularly fatuous exercise. They aren't doing an efficient policing job, nor are they meant to be; their prime role seems to consist in playing Aunt Sally to every idle youth who can lay his hand on a broken paving-stone or a gallon of petrol. As well, that is, as providing target practice for IRA gunmen (this applies particularly to soldiers afoot on the terrible 'target patrols'). If the Army were withdrawn to a central 'militarised zone' between, but not in, residential areas, where they could protect Catholic and Protestant gangs from each other, both communities might conceivably become more enchanted with them. As for policing in the Catholic areas—it is unfortunate, but still a fact of life, that the Protestant-dominated Royal Ulster Constabulary are unlikely to be accepted for some time. It is also a fact that no efficient force for the maintenance of law has ever been *imposed* on a community. The only interim solution would be *de facto* recognition and training of street vigilantes for local policing duties. There are precedents for this.[2]

Finally—how can a parent or teacher keep a child out of trouble on the streets? Can he be prevented from absorbing the violent standards of a violent society? Can he be protected from crippling anxiety?

A great deal was once made of setting a child 'an example'. Now we talk about 'providing behavioural cues', but the principle is exactly the same. What is more, it has turned out to be just as valid and as important as the Victorians said it was. Experiments in the psychological laboratory and the community, summarised in the last two chapters, have demonstrated the potency of such cues as angry words, violence observed and condoned, riot games and even toy guns. These play a large part in determining how a child will behave. We have seen too that this effect is at its maximum where an opportunity exists for the child to translate these cues into action.

Where this opportunity, in the shape of a warring community, cannot be removed, there is still positive action that can be taken. As the work in three Belfast schools has amply demonstrated, by excluding all cues to violence and substituting education in its derivative sense, children *can* be protected from the corrosive effects of violence on the streets.

This can of course be translated into parent–child terms. The most violent youngsters are those whose parents express impotence and anger with the regime.[1] However, such parents also feel it important that their children should support their own political stand, and there is little point in addressing to them any message that in their eyes would seem to undermine family solidarity. But the important point for all parents in this type of situation is that, in a crisis, the child looks for reaction cues from *them*; his behaviour will depend on the answers he gets to his unspoken questions. He may well make his own translation into more concrete terms, and an intolerant remark at home *could* have its final expression in a stone or petrol-bomb in the street. The implied order for a parent may be a tall one, but I am not making it. It is simply a logical extension from the evidence.

The same applies to cues to emotional response, the important factor being not the event itself, or even the amount it is talked about, but, as we have seen, the emotion conveyed by important adults. In the disturbed children seen both in Belfast and other disaster areas, the outstanding factor was that they were provided with cues to extreme anxiety. Conversely, effective treatment consisted in the child's reliving the experience, verbally or in play, with a therapist who provided reaction cues of a different kind. After the California earthquake, a psychiatrist dealt very effectively with groups of disturbed children by encouraging them to out-do each other by boasting about their experiences. Not necessarily the only way of handling this type of problem, perhaps, but an excellent example of the way in which, with ingenuity, a child's outlook and emotional response can be radically altered.

The earthquake had, of course, come and gone—all in a matter of hours. But can the brief, rather technical phrase 'providing response cues' really contain a philosophy of treatment to cope with a disaster that goes on and on?

What could a parent or doctor say, for example, to Mary? She is eleven and lives in a street that is still invaded night after night by gunmen and combat troops. Gunfire and bursting nail-bombs keep her awake until the small hours. Her brother is interned and her father, a vigilante, is engaged in sporadic battles with the security forces, sometimes verbal, sometimes physical. She comes to the doctor because of gross anxiety, nightmares and bed-wetting; at night, half-awake, she screams out her fears that her father will be shot or interned, or their house burnt down. All this, the doctor is unhappily aware, is quite likely.

What can he say to Seamus, aged thirteen, whose brother was blown to pieces while making a bomb, who has severe nightmares and school phobia, and who wants to join the IRA? Angry tears well up as he speaks of his resolve to drive the British 'invaders' out of his street and out of his country. Or what about Bill, the son of a member of the Ulster Defence Regiment? After a spate of fireside assassinations of UDR officers, he collapses with choking fits every time the doorbell rings.

These are certainly unpromising situations, to say the least; they confront us near the end of what must seem a very pessimistic book. But it is possible, even now, to set two positive and very encouraging factors against the dismal catalogue of fears, symptoms and aggressions that have resulted from five years of street violence. The first, based on a point made a couple of pages back, is that a child *can*, to some extent, be prevented from absorbing the brutal kill-or-be-killed values of civil war, even when they pervade his city and his street. The second is that children *can* be protected from emotional illness, even in an arena of on-going violence with its very real risks of injury and separation. Fortunately, the concept of therapy for the children described above, and many more, is not altogether bizarre, because most of the answers come, not from the children who break down emotionally, but from those who do not. The average child under stress is surprisingly resilient—and this is because his resilience is not all his own. It exists largely as a quality, or is lacking as a deficiency, in the adults who are closest to him. In fact, it is sometimes helpful to think of a child as experiencing a frightening event *at second hand*; just as with aggressive behaviour, his response is modelled on adult cues, rather than formed by direct reaction to the event itself.

Even if this is true, it may seem unhelpful in practical terms. Is it not even more difficult to treat a depressed or anxious adult who has, if anything, an even clearer idea of the future's frightening possibilities? The work described in Chapter 4, of course, strongly suggests that this is the case. But in the event, it appears that a large proportion of childhood symptoms in the context of violence or disaster can be attributed to basic handling misconceptions. For example, a great deal of emphasis is often laid on 'verbalisation' of anxiety—the notion that anxiety can be dispelled by the simple act of a child's talking about his fears, or through 'anxiety play'—building barricades, drawing pictures of armed soldiers and of burning houses. The *depletion* metaphor is just as tempting here as in the case of aggression (Chapter 8)—and just as misleading and potentially damaging. (Indeed, I offer here my belief that all simple mechanistic concepts in psychiatry are to be deeply distrusted.)

I am suggesting, in short, that verbalisation of anxiety has little or no value in itself. It is quite true that, immediately after a period of acute stress or a disaster, a child has a great need to talk about and to discuss his experiences—but that is in order that he can receive reassurance and support from adults or other children who have shared the experience; there is no evidence that the act of putting the experience into words has in itself any cathartic value—perhaps the reverse. At least one form of reiteration, in fantasy, of a traumatic event had extremely unfortunate effects. On the day that has come to be called 'Bloody Friday', when bombs went off all over Belfast, killing and mutilating shoppers, the TV newsreels ran the story over and over again—a fireman lifting a severed head from the pavement, arms and legs in polythene bags. Many viewers, both adults and children, were physically sick—and several children whom I knew of had severe night terrors for up to a week afterwards. In his school-classroom, one ten-year-old boy spent an entire day sitting ashen-faced, compulsively tapping his hand on his knee. Broadcasters may claim that the terrible realism of these news programmes will provoke public revulsion against terrorism, but this unproven advantage has to be weighed against real damage to children's mental health. In general, talk, fantasy and preoccupation arising from a frightening event is much more productive of anxiety than the event itself. There is a parallel here with the fact, noted in Chapter 4, that people living in areas where trouble was expected were significantly more prone to mental illness than those living in the actual troubled areas. In our own child psychiatry in-patient unit, the TV set is turned off at news-time, and this by general consent.

Escapism? Cloud-cuckoo land? I have been publicly accused at least once of being quite unrealistic.[3] So has the Belfast headmaster to whom I referred on p. 132, and whose report is, I think, worth quoting again at this point:

'We banned all reference to what was going on outside and as a result have created our own "no-go land" where they still read fairy stories, work out their fantasies in plays about Cinderella and so on. Unreal you may think, but I assure you that I have a school full of happy children who arrive shortly after 8 a.m. each morning, never leave the grounds all day and have to be beaten out of the place when it is closed up at 5 p.m.

'We have no neurotics and no problem children and an average attendance of over 90%. We do have problem parents and problem teachers—but problem children, no.'[4]

Like this headmaster, I can see no reason why bloody violence should be dangled constantly before children's eyes just because it happens to be taking place in the next town, or the next street. The reality exists and is ultimately inescapable, but I believe it to be a central principle that it should colour as few areas of a child's life as humanly possible. Perhaps it is not to be entirely crowded out, but the attempt seems to me eminently worthwhile. This approach, opening as it does new perspectives beyond a drab and violent environment, has been found most rewarding in several settings; it has made children less, rather than more, vulnerable.[5]

But what about the child who, understandably, *is* anxious when his street is in uproar, or who develops symptoms later and who, as part (but only as part) of his treatment, recapitulates his experiences verbally or in fantasy? When a situation is clearly deteriorating, what is the value of a calm, even a rigidly controlled, adult exterior; what is the use of phrases like 'Don't worry—you'll be okay—everything's going to be all right'?

Reassurance does not, fortunately, depend on a successful appeal to reason. If it did, we would all go in perpetual dread of the inevitable personal tragedy. But children share the general tendency of humans in a frightening situation to read any message, verbal, or non-verbal, as the one they had hoped for. And children are not, I think, quite as perceptive as is generally believed. For a child, and even for an injured and frightened adult, the value of a calm approach and of optimistic phrases consists purely in the emotional atmosphere they create. They are, simply, response cues . . .

On 'Bloody Friday' Tony, aged eight, was in town with his mother—like thousands of other shoppers. Panic took hold of the crowd as bombs started to go off, now in one street, now in another; people rushed here and there in search of an elusive sanctuary. Tony was brought home in a hysterical condition, screaming and shaking, and required heavy sedation from his GP. I saw him two weeks later. Ever since the day of the bombs he had suffered from chronic and gross anxiety; he had bowel and stomach symptoms and frequent night terrors in which the theme of sudden death constantly recurred. During the day he had fainting fits, and had to be removed from school.

In hospital, detailed tests excluded any physical cause for his symptoms and, as a last resort, his mother took him to Italy for some weeks in the hope that he would forget his traumatic experience. Unfortunately his symptoms persisted and worsened; after a premature return home he had to be admitted to our in-patient unit. Although the unit is located on the outskirts of Belfast, Tony's symptoms disappeared almost at once.

Examination of his family situation helps to explain this rather anomalous recovery. Tony is the eldest of five children; his father had died, after a long illness, over a year previously. His mother, a very dependent and talkative woman, still felt the loss greatly, and was oppressed with anxiety about the family shop and the general situation. A great believer in 'ventilation', she continually talked about her worries and Tony, as the eldest, had become her emotional prop and was daily in receipt of a burden of anxiety which he could not possibly have had the resources to cope with. In hospital, with guidance, he was very soon able to talk about his experience with the bombs—and in due course to dismiss it—without any accompanying symptoms of anxiety. This, of course, was only partial treatment. It remained for a skilful social worker to guide Tony's mother towards an understanding of the real stresses that had led to his breakdown and also for the same social worker to provide for the mother some of the emotional support she had previously sought from Tony. She was also able to learn a great deal by observing, and working with, a nurse-therapist who, during Tony's anxious periods, was adept at channelling his interest into creative activity. Tony went out of hospital for several weekends without symptoms; a cheerful, rather cheeky little boy, he was then ready for discharge.

Bill, the UDR officer's son, had also to be admitted to hospital. His father believed that his ten-year-old son should be 'tough' and had insisted on discussing with him his work and the risks involved, in detail. However, he was himself an emotionally vulnerable man, whose severe anxiety about his role in security and his family's physical safety had on several occasions almost led him to resign his commission. Bill's symptoms cleared, too, on admission; the difficulty lay in convincing his father that he was making demands on the child beyond his emotional capability. He still cannot quite accept Bill's natural need for support and reassurance. But, almost in spite of himself, he has learned a great deal in informal joint interviews with Bill and the nurse-therapist assigned to him (on similar lines to the joint interviews in Tony's case), and Bill will shortly be ready for discharge.

Bill's age is significant; in fact, all of the children we have seen with severe emotional symptoms resulting from the street violence have been aged about ten. A child of this age has a lively comprehension of the dangers of a situation, while at the same time remaining very dependent on adult reaction cues.

It will never be possible to protect a child from frightening experiences. Indeed, the facts about what is happening should not be withheld from a curious child, whose fantasies are likely to depict the events as much worse than they really are. But in the last analysis, it is largely the nature of the adult response, rather than any characteristics of the event itself, that will either harm or protect him.

Could this become a losing battle? What if the day came when no fantasy could do justice to a gory reality? If this point were to be reached, and it yet might be, then the battle would almost certainly be lost. So far, however, these protective methods have proved their value. And, as I have said, no doctor despises palliatives. For a State, just as for a patient, they can at the very least ensure a painless death.

II

References

Chapter 1: Fact or Fantasy?
1 The Provisional IRA. See Chapter 3.
2 Junior membership of the Orange Order. See Chapter 3.
3 In many areas, all the street lamps are broken, and the windows of houses and shops firmly shuttered when darkness falls. So that, from the well-lit main thoroughfares, the side-streets appear as black, terrifying chasms.
4 9 September 1971 brought a statement from Belfast Corporation that, as the traffic lights on the Falls Road had been repaired fifteen times since the end of 1969, and were once more damaged and out of action, they would now be left in their present condition.
5 See, for example, *New York Times*, 14 March 1971; *Newsweek*, 5 April and 19 April 1971; *Daily Express*, 10 December 1971.

Chapter 2: The Matrix
1 See also: Friedman, P., 'The Jewish Ghettos of the Nazi Era', *Jewish Social Studies*, Vol.xvi, No.1, 1954
   Schoenberger, G., *The Yellow Star*, Corgi Books, 1969
   Zangwill, I., *Children of the Ghetto*, 1894
2 Or, inversely, the 'solution' imposed on indigenes by prosperous settlers, as in South Africa. (Desmond, C., *The Discarded People*, Penguin Books, 1971. Contrast *As they came—in Africa*, United Nations, 1971.)
3 Evans, E.E., 'Belfast—the Site and the City', *Ulster Journal of Archaeologists*, 1944 (Third series), Vol.7, pp.25-9.
   Jones, E., *A Social Geography of Belfast*, London, 1960

Beckett, J.C., and Glasscock, R.E., 'Belfast: Origin and Growth of an Industrial City', BBC, 1967

4 Boyd, A., *Holy War in Belfast*, Anvil, 1969

5 Evans, E.E., in Beckett and Glasscock, *op. cit.* (Ch.1)

6 Boserup, A., and Iversen, C., 'Rank Analysis of a Polarised Community: A Case Study from Northern Ireland', Peace Research Society (*International Papers*), Vol.7, pp.59-76.

Boal, F.W., 'Territoriality on the Shankill-Falls Divide, Belfast', *Irish Geography*, 1969, Vol.6, No.1.

Boal, F.W., 'Social Space in the Belfast Urban Area', *Irish Geographical Studies*, 1970, Dept. of Geography, Queen's University, Belfast

7 Not an easy group to define in concrete terms. 'Low-income groups' is a fair alternative. See Ch.1 in Hoggart (Note 8).

8 Hoggart, R., *The Uses of Literacy*, Chatto & Windus, London, 1957

9 Horne, D., *God is an Englishman*, Penguin Books, Harmondsworth, 1969

10 *Belfast Telegraph*, 27 March 1971

Belfast Newsletter, 14 April 1971

Irish News, 14 April 1971

Protestant Telegraph, 27 March 1971

11 *Report of Northern Ireland Boundaries Commission*, 15 September 1971

12 Hoggart, R., *op. cit.*

13 'Children in Flats: A Family Study', NSPCC, London, 1970. An excellent bibliography is included.

14 cf. *St Luke*, Ch.19, v.40.

15 For example, one of the spurious 'Prophecies of Columcille' was that Ulster would finally be conquered by a foreign knight on a white horse —actually a legend with its roots deep in Irish folklore. In the reign of Henry II, one John de Courcy identified with this hero to the extent of believing that he was the man named. Having built a castle at Carrickfergus (still a landmark), he set out on a white horse to prove his claim, failed disastrously, and went on the Crusades instead.

16 A slang term for Catholics derived from the Fenian Society of the nineteenth century, an organisation dedicated to Irish independence. The word is much favoured by Protestants, since it is usually preceded by an adjective with the same initial letter. (A useful study might well concern itself with the use of plosive consonants in terms of cross-cultural abuse.)

17 More correctly, the 'Union Flag', but the more popular term is used.

18 Unilateral Declaration of Independence, a phrase which became current in Britain at the time of the Rhodesian crisis.

19 Mallory, L., *New Spectator*, August 1971

20 'Which are the Hooligans, Mr Hume?' Ed. in *Voice of the North*, Belfast, 24 October 1971

    Kupfer, M., *Newsweek*, 5 April 1971

    'Law(?) and Orders', Central Citizens' Defence Committee, Belfast, 1970

21 Rose, R., *Governing without Consensus*, Faber and Faber, London, 1971

22 *Belfast Newsletter*, 19 August 1971

23 Boal, F.W., *op. cit.*

24 Jackson, H., *The Two Irelands: A Dual Study of Inter-Group Tensions*, Minority Rights Group, Report No.2, 1970

25 Conway, Cardinal William, *Catholic Schools*, Catholic Communications Institute of Ireland, 1970

26 *Flight*, Northern Ireland Community Relations Commission, September 1971

27 Clare, J., *The Times*, 18 September 1971

28 The monthly Northern Ireland unemployment figures are about 45,000 (one male in ten being unemployed), the highest for some 30 years. The proportion unemployed in the 'ghetto' areas is, of course, much higher. No official figures are available, but in Ballymurphy, a Catholic working-class district, a recent estimate put the percentage of unemployed fathers at 47.

29 The Child Poverty Action Group has received 'disturbing reports of desperate poverty' in Belfast, and is conducting a detailed inquiry. (Press release, 24 November 1971; Townsend, Professor P. (Chairman), personal communication, 23 November 1971.)

Chapter 3: Papers and Preachers

1 Frith, G., *Fortnight*, 14 May 1972

2 Allport, G.W., *The Nature of Prejudice*, Addison-Wesley, Reading, Mass., 1954

3 Having had no tea, Alice could not therefore be offered 'more'. (Lewis Carroll, *Alice in Wonderland*, Ch.7.) Demands by Nationalist groups for 'a return to a united Ireland' have the same fairy-tale quality. Ireland has never been united—except, uneasily, under British rule.

4 *Protestant Telegraph*, Vol.4, No.7

5 *Ibid.*, 5, 18

6 *Ibid.*, 5, 18

7 *Ibid.*, 5, 17

8 *Ibid.*, 6, 7

9 See also: Huxley, Aldous, *Brave New World Revisited* (Ch.7), Chatto & Windus, London, 1950

Sargant, W., *Battle for the Mind*, Heinemann, London, and Doubleday, New York, 1957

Wesley, J., *Journal*, Isbister, London, 1902

10 Dewar, M. W. *et al.*, *Orangeism: A New Historical Appreciation*, Grand Orange Lodge of Ireland, 1967

11 Jackson, H., *The Two Irelands: A dual Study of Inter-Group Tensions*, Minority Rights Group, Report No.2, 1970

12 *Irish News*, 19 April 1971

13 *Daily Star*, Lebanon, 2 April 1972; *This Week*, Dublin, 6 July 1972; *Belfast Newsletter*, 7 September 1972; *Daily Express*, 12 September 1972

14 See also: Coogan, T.P., *The I.R.A.*, Collins, London, 1970

Moss, R., *Urban Guerrillas* (Ch.4), Temple Smith, London, 1972

Sunday Times Insight Team, *Ulster*, Penguin Books, Harmondsworth, 1972

15 I do not, of course, claim for my own account any advantage over the many others available – except perhaps brevity. See:

Disturbances in Northern Ireland (Cameron Report), Cmd. 532, HMSO, 1969

Violence and Civil Disturbances in Northern Ireland in 1969 (Scarman Report), Cmd. 566, HMSO, 1972

Sunday Times Times Insight Team, *op. cit.*

Egan, B., and McCormack, V., *Burntollet*, LRS Publishers, Belfast, 1969

Devlin, B., *The Price of my Soul*, André Deutsch, London, 1969

Wallace, M., *Northern Ireland: 50 Years of Self Government*, David and Charles, Newton Abbott, 1971

Chapter 4: The Cost of commotion: 1969 Onwards

1 Royal Ulster Constabulary Statistics, September 1971

2 Clarke, P., *Guardian*, 23 August 1971

3 *Observer*, 29 August 1971

4 *Belfast Telegraph*, 7 September 1971

5 *Belfast Newsletter*, 2 September 1971

6 Sinclair, F., *Belfast Telegraph*, 16 September 1971

7 Clenaghan, A.S., Reported *Belfast Newsletter*, 29 March 1971

8 Thompson, J.F., Personal communication, 17 September 1971

9 Lyons, H.A., 'Psychiatric Sequelae of the Belfast Riots', *British Journal of Psychiatry*, 1971, Vol.118, pp.256-73

10 Legrand, du Saulle, H., 'De l'État Mental des Habitants de Paris pendant les événements de 1870–71', *Annales Medico-Psychologiques*, 1871, Vol.ii, pp.222–41

11 Emslie, I., 'The War and Psychiatry', *Edinburgh Medical Journal*, 1915, Vol.14, pp.359–67

12 Mira, A., 'Psychiatric Experience in the Spanish Civil War', *British Medical Journal*, 1939, Vol.i, pp.1217-20

13 Brown, E., *Hansard*, 30 June 1942

14 Lewis, A., 'Incidence of Neurosis in England under War Conditions', *Lancet*, 1941, Vol.ii, pp.175-83

15 Atkin, T., 'Air-Raid Strain in Mental Hospital Admissions', *Lancet*, 1941, Vol.ii, pp.72-4

16 Harris, A., 'Psychiatric reactions of Civilians in War-Time', 1941

17 Pegg, G., 'Psychiatric Casualties in London, September 1940', *British Medical Journal*, 1940, Vol.ii, pp.553-5

18 Massey, A., report of meeting at Tavistock Clinic, *British Medical Journal*, 1941, Vol.i, p.77

19 Hemphill, R.E., 'The Influence of War on Mental Disease', *Journal of Mental Science*, 1941, Vol.87, pp.170-82

20 Brown, F., 'Civilian Psychiatric Air-Raid Casualties', *Lancet*, 1941, Vol.i, pp.686-91

21 Harrisson, T., *British Medical Journal*, 1941, Vol.ii, pp.573 and 832

22 Allen, C., *British Medical Journal*, 1941, Vol.ii, p.961

23 Ødegård, Ø., 'The Incidence of Mental Diseases in Norway during World War II', *Acta Psychiatrica et Neurologica Scandinavica*, 1954, Vol.29, pp.333-53

24 Dohan, F.C., 'Wartime Changes in Hospital Admissions for Schizophrenia', *Acta Psychiatrica Scandinavica*, 1966, Vol.42, pp.1-23

25 Murney, H., *Statistical Report on the Injuries Sustained during the Riots in Belfast from 8th to 22nd August, 1864*, Belfast, 1864

26 Lyons, A., *op.cit.*

27 Bodman, F., 'War Conditions and the Mental Health of the Child', *British Medical Journal*, 1941, Vol.ii, pp.486-8

28 Mons, W.E.R., 'Air Raids and the Child', *Lancet*, 1941, Vol.ii, pp.625-6

29 When a difference is said to be 'statistically significant', it means that it is highly unlikely to be due to chance alone. *How* unlikely this is is expressed as a percentage: thus, if a difference is said to be significant at the 5% level of probability, it means that there are less than five chances in a hundred that the difference is only a chance one. This (5%) is the level of significance generally accepted in scientific research.

30 Mental illness is usually divided into two broad categories: the psychoses and the neuroses. The former is the more severe type of disorder, where contact with reality is lost; reality is distorted or replaced with fantasy. Schizophrenia is the largest single group in this category.

The neurotic illnesses comprise depressive reactions and anxiety states; contact with reality is maintained, but the patient responds to stress with psychiatric symptoms rather than with useful, adaptive action.

31 All drugs used for daytime sedation are here termed tranquillisers, and those for night sedation hypnotics. Anti-depressants are used for the purpose that their name suggests.

32 After the blast at the Europa Hotel, Belfast, 2 September 1971

33 Brown, G.W., and Birley, J.L.T., 'Crises and Life Changes and the Onset of Schizophrenia', *Journal of Health and Social Behaviour*, 1968, Vol.9, pp.203-14

34 Birley, J.L.T., and Brown, G.W., 'Crises and Life Changes preceding the Onset or Relapse of Acute Schizophrenia: Clinical Aspects', *British Journal of Psychiatry*, 1970, Vol.116, pp.327-33

Chapter 6: Disorders and Defences

1 Dunlap *et al.*, 'Young Children and the Watts Revolt', Southern California Permanente Medical Group, 1966

2 Marks, I.M., and Gelder, M.G., 'Different Ages of Onset in Varieties of Phobia', *American Journal of Psychiatry*, 1966, Vol.123, pp.218-21

3 Roth, M., 'The Phobic Anxiety-Depersonalization Syndrome', *Proceedings of the Royal Society of Medicine*, 1939, Vol.52, p.587

4 Shorvon, H.J., 'The Depersonalization Syndrome', *Proceedings of the Royal Society of Medicine*, 1946, Vol.39, p.776

5 Burton, L., *Vulnerable Children*, Routledge and Kegan Paul, London, 1968

6 Bodman, F., 'War Conditions and the Mental Health of the Child', *British Medical Journal*, 1941, Vol.ii, pp.486-8

7 Burbury, W.M., 'Effects of Evacuation and of Air-Raids on City Children', *British Medical Journal*, 1941, Vol.ii, pp.660-2

8 Mons, W.E.M., 'Air Raids and the Child', *Lancet*, 1941, Vol.ii, pp.625-6

9 Bowlby, J., *Maternal Care and Mental Health*, World Health Organisation, 1952

10 Freud, A., and Burlingham, D., *War and Children*, Medical War Books, New York, 1943

11 Schwebel, M., 'Studies of Children's Reactions to the Atomic Threat',

American Journal of Orthopsychiatry, 1963, Vol.33, pp.202–3

12 Darr, J.W., 'The Impact of the Nuclear Threat on Children', *American Journal of Orthopsychiatry*, 1963, Vol.33, pp.203–4.

13 Escalona, S.K., 'Children's Awareness of the Threat of War—Some Developmental Implications', *American Journal of Orthopsychiatry*, 1963, Vol.33, pp.204-5

14 Adams, J.F., 'Adolescent Opinion on National Problems', *Personnel Guidance Journal*, 1963, Vol.42(4), pp.397-400

15 Allerhand, M.E., 'Children's Reactions to Societal Crises', *American Journal of Orthopsychiatry*, 1965, Vol.35, pp.124-30

16 Elder, J.H., 'A Summary of Research on Reactions of Children to Nuclear War', *American Journal of Orthopsychiatry*, 1965, Vol.35, pp.120-3

17 Wrightsman, L.S., 'Parental Attitudes and Behaviors as Determinants of Children's Responses to the Threat of Nuclear War', paper presented at American Psychological Association, Philadelphia, 1963

18 Nemtsow, J., and Lesser, S.R., 'Reactions of Children and Parents to the Death of President Kennedy', *American Journal of Orthopsychiatry*, 1964, Vol.34, pp.280-1

19 Sigal, R., 'Child and Adult Reactions to the Assassination of President Kennedy', paper read at American Psychological Association, Los Angeles, 1964

20 Eiduson, B.T., 'A Study of Children's Attitudes to the Cuban Crisis', *Mental Hygiene*, 1965, Vol.49, p.113

21 Perry, S.E. *et al.*, 'The Child and his Family in Disasters: A Study of the 1953 Vicksburg Tornado', *NAS-NRC*, No.394, Washington, D.C., 1956

22 Crawshaw, R., 'Reactions to a Disaster', *Archives of General Psychiatry*, 1963, Vol.9, pp.157-62

23 Brussilowski, L., 'Neuropsychiatric Reactions to Earthquakes', *Ztschr. F.D. ges. Neurol. u. Psychiat.*, 1928, pp.116-442

24 Hansen, H., 'Psychiatric Aspects of Hospitalized Children in Disaster; the California Earthquake, 1971', paper read at American Association of Psychiatric Services for Children, Los Angeles, 1971

25 San Fernando Valley Child Guidance Clinic, 'Mental Health Implications in a Disaster', seminar, Los Angeles, 1971

26 Howard, S.J., paper read at American Association of Psychiatric Services for Children, Los Angeles, 1971

27 Dunlap *et al.*, *op. cit.*

28 Berry, M., MD personal communication

29 Cooper, P., and Branthwaite, A., Manchester University, 1969

30 Hansen, H., *op. cit.*

31 Bambridge, G., *Daily Telegraph*, 25 March 1971

32 Allerton, W.S., 'Mass Casualty Care and Human Behavior', Medical Annals D.C., 1961, Vol.33, pp.206-8

33 Crawshaw, R., *op. cit.*

34 Friedman, P., and Lewis, L., 'Some Psychiatric Notes on the Andrea Doria Disaster', *Journal of the American Medical Association*, 1959, Vol.171, p.222

35 Fritz, C.E., and Williams, H.B., 'The Human Being in Disaster', *Annals of the American Academy of Political and Social Sciences*, 1957, Vol.309, pp.42-51

36 Glass, A.J., 'Management of Mass Psychiatric Casualties', *Military Medicine*, 1956, Vol.118, pp.335-41

37 McGonagle, L.C., 'Psychological Aspects of Disaster', *American Journal of Public Health*, 1964, Vol.54, pp.638-43

38 Popovic, M., and Petrovic, D., 'After the Earthquake', *Lancet*, 1964, pp.1169-71

39 Robinson, D., *The Face of Disaster*, Doubleday, New York, 1959

40 Sigurdson, W.E., 'The Psychiatric Aspects of Mass Disaster', *Canadian Journal of Public Health*, 1958, Vol.49, pp.288-91

41 *Belfast Telegraph*, 12 August 1971

42 *Ulster Protestant*, Belfast, October 1971

43 McCarthy, P., personal communication

Chapter 7: Republicans and Sinners

1 Target, G., *Unholy Smoke*, Hodder & Stoughton, London, 1969

2 The terms 'Celtic' and 'Anglo-Saxon' are not strictly accurate, but are as near as one can get without lengthy *qualifications*.

3 *Second Report of N.I. Commissioner for Complaints*, 3 November 1970

4 de Molinari, Gustave, *Journal des Débats*, 1880. Quoted in a *Times* leader of 18 September 1880, and by Lewis P. Curtis in *Apes and Angels* (David and Charles, Newton Abbott, 1971), to whom I am indebted for other valuable insights into Victorian stereotype.

5 Curtis, Lewis P., *op. cit.*

6 Barbu, Z., 'Nationalism as a Source of Aggression', in *Conflict in Society*, Ciba Foundation, 1966

7 Jaench, E.R., *Der Gegentypus*, Barth, Leipzig, 1938

8 For further reading, see: Gunterman, 'A Bibliography on Violence and Social Change', in *Urban Riots*, ed. Robert H. Connery, Vintage Books, New York, 1969

9 Lennon, P., *Sunday Times*, 12 September 1971

10 Eysenck, H.J., *Uses and Abuses of Psychology*, Penguin Books, Harmondsworth, 1953

11 Daniel, W.W., *Racial Discrimination in England*, PEP, 1968

12 Gardner, L., *Resurgence of the Majority*, Ulster Vanguard Publications, 1971. The Orange Order's Grand Master for Belfast, in a foreword, commends the work as a 'mine of information'.

13 *Protestant Telegraph*, 16 January 1971

14 Barritt, D.P., and Carter, C.F., *The Northern Ireland Problem*, London, 1968

15 Rose, R., *Governing without Consensus*, Faber and Faber, London, 1971

16 Allport, G.W., *The Nature of Prejudice*, Addison-Wesley, Reading, Mass., 1954

17 Bettleheim, B., and Janowitz, M., *Dynamics of Prejudice*, Harper & Row, New York, 1950

18 Campbell, in Allport, G.W., *op. cit.*

19 For a detailed review, see Berkowitz, L., *Aggression: A Social Psychological Analysis*, McGraw-Hill, New York, 1962. See also: White, R.K., and Lippitt, R., *Autocracy and Democracy: An Experimental Inquiry*, Harper & Row, New York, 1960

20 Berkowitz, L., *op. cit.*

21 *Behind Convent Walls*, 'ex-nun', Protestant Truth Society

22 *Protestant Telegraph*, 30 October 1971

23 Hashmi, F., in *Because They're Black*, Humphry, D., and John, G., Penguin Books, Harmondsworth, 1971

24 See Chapter 3.

25 Quoted in *Newsweek*, 19 April 1972

26 *Punch*, 1906, Vol.130, p.460

27 Peters, W., *A Class Divided*, Doubleday, New York, 1971

28 Bayley, J., and Loizos, P., 'Bogside off its Knees', *New Society*, 21 October 1969

29 Boserup, A., *Revolution and Counter-Revolution in Northern Ireland*, Institute for Peace and Conflict Research, Denmark, 1970

30 Jenkins, R., and McCrae, J., *Religion, Conflict and Polarization in Northern Ireland*, Publication 5-4, Peace Research Centre, 1967

31 Mørch, Sveir, *On the Conflictual Relations of the Protestant and Catholic Groups in the Cities of Northern Ireland*, Institute for Peace and Conflict Research, Denmark, 1970

32 *Disturbances in Northern Ireland* (Cameron Report), HMSO, Belfast, 1969

33 Sunday Times Insight Team, *Ulster*, Penguin Books, Harmondsworth, 1972

34 Moss, R., in *Urban Guerrillas* (Temple Smith, 1972), for example, stresses the importance of this factor in the present-day drift to the cities of Latin America.

35 Clark, Kenneth B., *Dark Ghetto*, Harper & Row, New York, 1965

36 Rainworth, Lee, *Behind Ghetto Walls*, Aldine Publishing Company, Chicago, 1970

37 *Report of the National Advisory Council on Civil Disorders*, New York, 1969

38 *McCone Commission Report on Watts Riot*, California State Printing Office, 1965

39 Lee, A. McC., *Race Riot*, Dryden Press, New York, 1943

40 de Toqueville, A., *The Old Régime and the French Revolution*, Harper, New York, 1856

41 Davies, James C., 'Towards a Theory of Revolution', *American Sociological Review*, 1962, Vol.27, pp.5-8, 15–18

42 Moss, R., *op. cit.*

43 See also Grier and Cobbs, *Black Rage*, Basic Books, New York, 1965

44 Festinger, L., *A Theory of Cognitive Dissonance*, Northwestern University Press, Evanston, Illinois, 1957

45 For a really harrowing account of what it feels like to be at the receiving end of anti-Black aggression, read *Black Like Me* by John Howard Griffin (Collins, 1962). Griffin, a white man, darkened his skin by chemical means and travelled in the deep South.

46 Nielsen, S.L., 'Intergroup Conflict and Violence, Belfast, 1968–?' *Psychosocial Studies No. 4*, University of Bergen, Norway, 1971

47 Lumsden, M., *A Test of Cognitive Balance Theory in a Field Situation: A Factor Analytic Study of Perceptions in the Cyprus Conflict*, 1970

48 Lumsden, M., personal communications, 1971

Chapter 8: The Route from Fantasy

1 *Newsweek*, 19 April 1971; *Times Educational Supplement*, 20 August 1971

2 *The Child's Guardian*, NSPCC, March 1971

3 Dollard, J.L. *et al.*, *Frustration and Aggression*, Yale University Press, New Haven, 1939

4 Berkowitz, L., *Aggression: A social Psychological Analysis*, McGraw-Hill, New York, 1962. This book includes a detailed review, with bibliography, of work on the frustration–aggression hypothesis.

5 *Romans*, Ch.7, v.22–5

6 Russell, James L., *Civic Education in Northern Ireland*, Report to The Northern Ireland Community Relations Commission, 1972. Russell found that boys who were members of gangs were more likely to approve of violence; within the gangs, the leaders were *most* likely to approve of violence.

7 A study at present in progress is likely, when completed, to show this same 'significant pre-occupation with death and destruction' among Belfast children between the ages of six and eight (Professor Rhona M. Fields, San Fernando Valley State College, personal communication).

8 Schmidt, H.D., 'Bigotry in Schoolchildren', *Commentary*, 1960, Vol.29, pp.253-7

9 Harding, J. *et al.*, 'Prejudice and Ethnic Relations', in G. Lindzey (Ed.), *Handbook of Social Psychology*, Addison-Wesley, Reading, Mass., 1954

10 Deutsch, M., and Collins, M.E., *Interracial Housing: A psychological Evaluation of a Social Experiment*, University of Minnesota Press, Minneapolis, 1951

11 Mussen, P.H., 'Some Personality and Social Factors Related to Changes in Children's Attitudes Towards Negroes', *Journal of Abnormal and Social Psychology*, 1950, Vol.45, pp.423-41

12 Lane, M.B., 'Nursery Schools in the Service of Mental Health', in *The Mental Health of the Child*, National Institute of Mental Health, 1971

13 Lundberg, G.A., and Dickson, L., 'Selective Association among Ethnic Groups in a High School Population', *American Sociology Review*, 1952, Vol.17, pp.23-5

14 Curtis, Lewis P., *Apes and Angels*, David and Charles, Newton Abbott, 1971

15 For example, in *The Weekly Freeman*, 4 December 1886. Cited in Curtis, Lewis P. (*op. cit.*)

16 *Loyalist News*, 28 August 1971

17 *The Ulster Protestant*, October 1971

18 *Protestant Telegraph*, 2 October 1971

19 Bandura, A. *et al.*, 'Transmission of Aggression Through Imitation of Aggressive Models', *Journal of Abnormal and Social Psychology*, 1961, Vol.63, pp.575-82

20 Bandura, A. *et al.*, 'Imitation of Film-Mediated Aggressive Models', *Journal of Abnormal and Social Psychology*, 1963, Vol.66, pp.3-11

21 Hicks, D.J., 'Imitation and Retention of Film-Mediated Aggressive Peer and Adult Models', *Journal of Personality and Social Psychology*, 1965, Vol.2, pp.97-100

22 Siegel, A.E., 'Aggressive Behaviour of Young Children in the Absence of an Adult', *Child Development*, 1957, Vol.28, pp.371-8

23 Siegel, A.E. *et al.*, 'Permissiveness, Permission, and Aggression: The Effect of Adult Presence or Absence on Aggression in Children's Play', *Child Development*, 1959, Vol.30, pp.131-41

24 Levin, H., and Turgeon, V.F., 'The Influence of Mother's Presence on Children's Doll-Play Aggression', *Journal of Abnormal and Social Psychology*, 1957, Vol.55, pp.304-8

25 *Report of Illinois White House Conference on Children and Youth*, Illinois Commission on Children, May 1970

26 Berkowitz, L., and Le Page, A., 'Weapons as Aggression-Eliciting Stimuli', *Journal of Personality and Social Psychology*, 1967, Vol.7, pp.202-7

27 Berkowitz, L., *Roots of Aggression*, Atherton Press, New York, 1969

28 *The Memoirs of General Grivas*, ed. Charles Foley, Longman, London, 1964

29 Maurer, A., 'Maturation of Concepts of Death', *Journal of Genetic Psychology*, 1964, Vol.5, p.105. See also bibliography appended to this paper, and: Anthony, S., *The Child's Discovery of Death*, Kegan Paul, London, 1940

30 Himmelweit, H.T. *et al.*, *Television and the Child*, Oxford University Press, 1958

31 *Action for Children's Television*, Report of the first National Symposium on Children and Television, October 1970

32 Schramm *et al.*, *Television in the Lives of our Children*, Stanford University Press, 1961

33 Bandura, A., 'The Impact of Visual Media on Personality', *The Mental Health of the Child*, National Institute of Mental Health, June 1971. See also detailed bibliography which is appended.

34 *Poetics*, vi, 1449

35 Lumsden, M., 'The Instinct of Aggression: Science or Ideology?', *Futurum*, March 1970

36 'we but teach Bloody instructions, which being taught return To plague the inventor.' *Macbeth*, Act I, Sc.vii

37 Murphy, M., *Belfast Telegraph*, 8 April 1971

38 Seth, Prof. G., and Caldwell, C., *Newsweek*, 19 April 1971

39 McElroy, J., *Times Educational Supplement*, 20 August 1971

40 Maultsaid, J. *et al.*, *Observer Review*, 12 December 1971

41 Connolly, S., *Belfast Telegraph*, 6 March 1972

Chapter 9: Education for Aggro

1 Spencer, A.E.C.W., *The Future of Catholic Education in England and Wales*, Catholic Renewal Movement, 1971

2 Corkey, W., *Episode in the History of Protestant Ulster, 1925–1947*, Belfast, 1960.

3 Long, Brother The Revd S.E., *Orangeism: A New Historical Appreciation*, Grand Orange Lodge of Ireland, 1967

4 Smyth, W.M., *The Battle for Northern Ireland*, Grand Orange Lodge of Ireland, 1972

5 Philbin, Dr W., quoted in Wallace, M., *Northern Ireland: 50 Years of Self-Government*, David and Charles, Newton Abbott, 1971

6 Dudgeon, J., *Irish News*, 8 March 1971

7 See Chapter 8, Refs. 1 and 40.

8 See literature reviews in Berkowitz, *op. cit.* (Chapter 8, Ref. 4) and Russell, *op. cit.* (Chapter 8, Ref. 6)

9 Tadjfel, L., 'Children and Foreigners', *New Society*, 30 June 1966

10 Goodman, M.E., *Race Awareness in Young Children*, Addison-Wesley, Reading, Mass., 1952

11 Magee, J., *The Teaching of Irish History in Irish Schools*, Northern Ireland Community Relations Commission, 1970

12 Barritt, D.P., and Carter, C.F., *The Northern Ireland Problem: A Study in Group Relations*, 1962

13 *Belfast Telegraph*, 10 December 1970

14 *WHERE*, May 1971

15 Rose, R., *Governing without Consensus*, Faber and Faber, London, 1971

16 Russell, J.L., *op. cit.* (Chapter 8, Ref. 6)

17 See Chapter 8, Ref. 4

18 Lane, M.B., 'Nursery Schools in the Service of Mental Health', *The Mental Health of the Child*, National Institute of Mental Health, June 1971

19 Lee, A.McC., *Race Riot*, Dryden Press, New York, 1943

20 Conway, Cardinal William, *Catholic Schools*, Catholic Communications Institute of Ireland, 1970

21 *Speak Out, No. 3 (N. Ireland)*, National Council for Civil Liberties, 1972

22 *Disturbances in Northern Ireland*, HMSO, 1969

23 *Belfast Newsletter*, 17 February 1972

24 *Draft Recommendation No. 7*, The Northern Ireland Alliance Party

25 Chambers, G. (N.I. Labour Party spokesman on education), 17 August

26 Longley, C., *The Times*, 25 February 1971

27 'Uneasy Jubilee', *Economist*, May 1971

28 *Belfast Telegraph*, 17 February, 18 February, 4 March 1971

29 *National Opinion Polls*, 1967

30 Burrows, P., *Community Forum*, Vol.1, No.1., Community Relations Commission, 1971

31 *The Student* (St. Malachy's Grammar School)

32 *The Irish Communist*, May 1971

33 Devlin, B., *The Price of My Soul*, Andre Deutsch, London, 1969

34 Quoted in Wallace, M., *op. cit.*

35 The unions for teachers in State and Catholic schools respectively.

36 Greeley, A.M. *et al.*, *The Social Effects of Catholic Education*, National Opinion Research Centre, Chicago, 1964
Greeley, A.M., and Rossi, P.H., *The Education of Catholic Americans*. Aldine Publishing Co., Chicago, 1966

38 Spencer, A.E.C.W., 'An Evaluation of Roman Catholic Educational Policy in England and Wales, 1900–1960', in Jebb, P., *Religious Education: Drift or Decision?*, Darton, Longman and Todd, London, 1968

39 *Violence and Civil Disturbances in Northern Ireland in 1969: Report of Tribunal of Inquiry*, HMSO, Cmd. 566

40 Jackson, B., and Marsden, D., *Education and the Working Class*, Routledge and Kegan Paul, London, 1962

Chapter 10: Prospect

1 Russell, James L., *op. cit.* (Chapter 8, Ref. 6)

2 The Ulster Special Constabulary, for example, was set up in 1920 to provide training, within a uniformed and disciplined framework, for members of Protestant vigilante-type groups. This was introduced as a short-term measure. In 1969 the B-Specials' crime, as indicted by Hunt (HMSO, Cmd. 535) lay in their having become an anachronism.

3 Nixon, K., *Belfast Newsletter*, 30 June 1972

4 Nihill, J.J., quoted in Bleakley, D., *Peace in Ulster*, Mowbray, Oxford, 1972

5 See also Fraser, R.M., 'At School During Guerrilla War', *Special Education*, Vol.61, No.2, June 1972